# DIY
# Solar Power

*The Ultimate Guide to Building a Home Photovoltaic System and Achieving Energy Self-Sufficiency*

by

**Ryan Greene**

Ryan Greene
*DIY SOLAR POWER*

# Table of Contents

# INTRODUCTION

Solar energy is, quite simply, the energy that comes from and is collected by the sun, and is the primary source of energy on the whole planet. It is a type of energy that runs on flow, not on stock, and therefore its continuous use does not diminish its availability in any way.

The ancient Greeks and Romans used mirrors to reflect light from the sun, concentrating its power to make fire and light torches. It is even claimed that the Greek scientist Archimedes used mirrors as a tool of war, concentrating the sun with enough strength to set fire to besieging Roman ships. These are examples of human ingenuity used to capture the power of the sun.

In addition, solar energy is the source of all other energies available in the world such as wind, fossil fuel, wave, hydroelectric, biomass, geothermal and tidal energy. It is a renewable, available, inexhaustible and clean source, thanks to the enormous amount of energy released by the sun every day. Solar energy is collected and used to produce electricity (through photovoltaic solar panels) or thermal energy (through solar thermal energy). There are three ways in which energy from the sun can be converted: solar photovoltaic, solar thermal and solar thermodynamic.

In this book we will obviously talk a lot about the electricity generated by solar energy and its uses, but we will deal with both concepts of solar electricity and passive solar, as both are useful for different types of projects and circumstances.

# CHAPTER 1: What are Solar Panels?

Solar panels transform solar energy into useful human energy, heat or electricity. Although similar on the outside, there are different solar panel technologies. The source of energy is always the same, solar energy, but some panels are useful for heating domestic water while others for the production of electricity.

## The types of solar panels

The main types of solar panels are the following:

**Photovoltaic solar panels.** These panels convert the sun's rays directly into electrical energy. The owner of the photovoltaic panels consumes self-produced electricity for domestic use or sells it to the national grid.

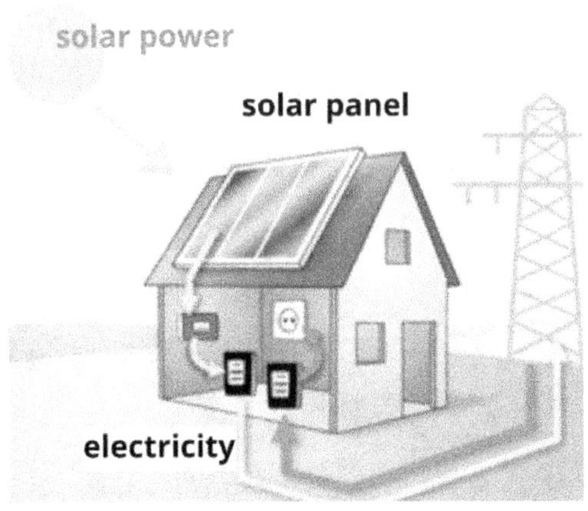

## Solar thermal panels (solar collectors)

They use the heat of the sun's rays to heat domestic water and produce hot water for domestic use in the bathroom and kitchen (e.g. washing dishes, showering, etc.). They are an ecological substitute for the electric water heater and gas boiler.

**Concentrated solar panels.** This technology uses parabolic mirrors to reflect the sun's rays and concentrate them in a single focal point that becomes particularly hot. This is the principle of Archimedes. The heat in the focal point can be used to heat a heat transfer fluid or a water tank, to cook food or to generate steam and thus electricity.

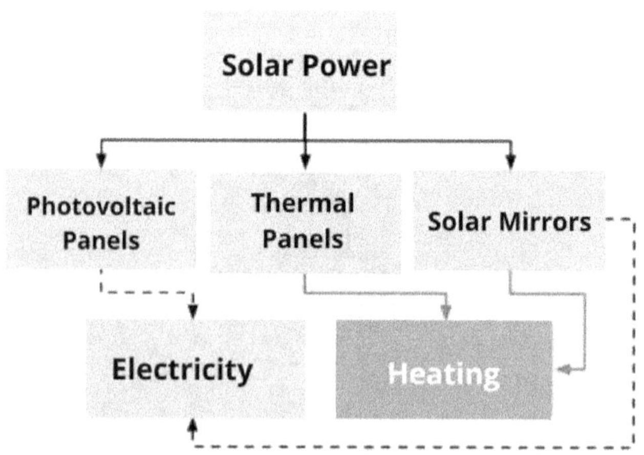

# CHAPTER 2: How the Solar Thermal Panel is Made and How it Works

The solar collector (or solar thermal panel) consists of a radiator/absorber - usually made of a metal such as copper, with good conduction capacity - which is able to absorb the heat of the sun's rays and transfer it to the water tank.

The operating principle of a solar thermal panel can be compared to what happens in a greenhouse. Of the incident sunlight, only a small part of the glass surface is reflected, the remaining part passes through the glass and is absorbed by a black capturing plate. This plate warms up and puts energy back into the form of infrared radiation, with respect to which the glass behaves as if it were opaque, thus keeping it inside (greenhouse effect). In this way the temperature of the primary vector fluid tends to heat up. From that moment the liquid moves in the coil towards the tank according to three different types of circulation: natural, forced or emptying.

The insolation depends on the cloudiness and the orientation of the panel with respect to the sun and a panel receives more solar energy when it is oriented directly towards the sun. The fixed panel provides the best yield when facing south and is inclined 10 degrees less than the latitude of the place if it has to produce hot water, 10 degrees more if it is used for heating. The United States is a country with a good level of radiation, on average 5-6 kwh/sqm/day.

## Types of Collector

The quality of the collector depends on the efficiency, i.e. the ability to convert solar energy into thermal energy. In technical terms, the efficiency of a solar collector is defined as the ratio between the energy (energy density) absorbed by the heat transfer fluid and the energy (solar energy density) incident on its surface.

With reference to the various types of panels and their efficiency we can summarize as follows.

| Type of panels | Price & Efficiency | Terms of use | Display |
|---|---|---|---|
| exposed. | Low | Ambient temperature of at least 20°C | Heating of outdoor swimming pools, hot water for showers bathing establishments, campsites, etc.. |
| glazed | Intermediate | March to October | Hot water for sanitary use |
| under vacuum. | High | Anything | Hot water for sanitary use and pre-heated for winter heating |

## Flat Glass Collector

The most used solar panel is the glass panel which is composed as follows:

1.  A sheet of transparent glass, placed above the absorber, which allows the passage of sunlight. The glass can withstand hail (normally tested with hail of 25mm diameter).

2.  Inside there is a solar light absorber, consisting of a sheet similar to a radiator (which may be made of steel or copper), inside which is inserted a bundle of tubes into which flows the liquid of the primary circuit intended to be heated. This fluid is normally water added with antifreeze in order to resist the winter cold without freezing.

The absorber, as it heats up, emits energy in the form of infrared radiation: but the glass, with respect to this radiation, attenuates the dispersion outside because it is opaque (greenhouse effect).

In the lower part of the panel there is a thermal insulation (in fibreglass or CFC-free polyurethane foam) that reduces heat loss.

The panel is closed at the rear by a shell, often made of sheet metal.

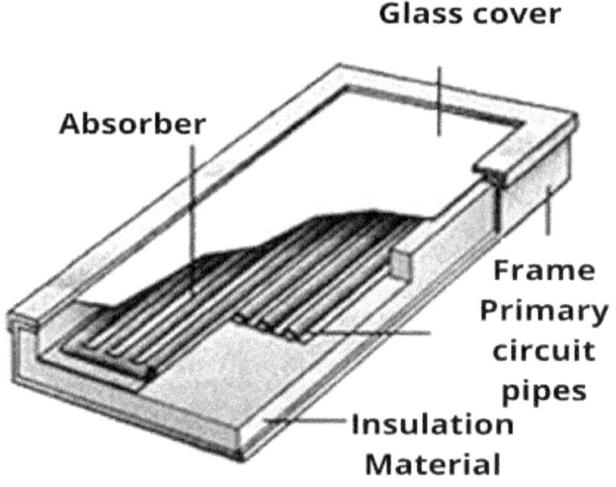


Thermally insulated casing

The whole (glass, absorber and tube bundle, thermal insulation and rear shell) is held together by a chassis that assembles the parts and gives the panel strength and stability.

## Non-selective glazed collector

Unlike the selective ones, the absorber is simply painted black. First-generation collectors consist of an insulated chamber between tempered glass directly exposed to the sun and an insulated rear shell. A blackened metal absorber is inserted inside the solar panel, so as to increase its efficiency.

## Selective plate glass solar collectors

The absorber undergoes an electro-chemical treatment under vacuum or by pigmentation in order to obtain a surface with a high absorption coefficient and a low reflection coefficient up to a temperature of 140°C.

The electrochemical treatment generally consists of a black chrome deposit on nickel, the vacuum treatment by means of titanium oxide vapours and the pigmentation treatment by means of metallic paints. The reduction of the reflection of the incident radiation allows the reduction of heat loss connected to it, therefore these panels have a higher efficiency than traditional panels, and have a cost about 10% higher than other panels. They have the advantage of being able to work better in colder and less sunny periods. For this reason, they can also be used in integration with the heating system.

## Uncovered panels (not glazed)

These panels have no cover, no insulation and no containment box. They are made of black plastic material (polypropylene, neoprene, PVC) to facilitate the absorption of heat and consist of a bundle of tubes inside which water passes through where it is heated directly by the sun's rays before being used.

They operate with an ambient temperature of at least 20°C (below the balance between stored and dispersed energy is unfavourable), and the maximum water temperature does not exceed 40°C.

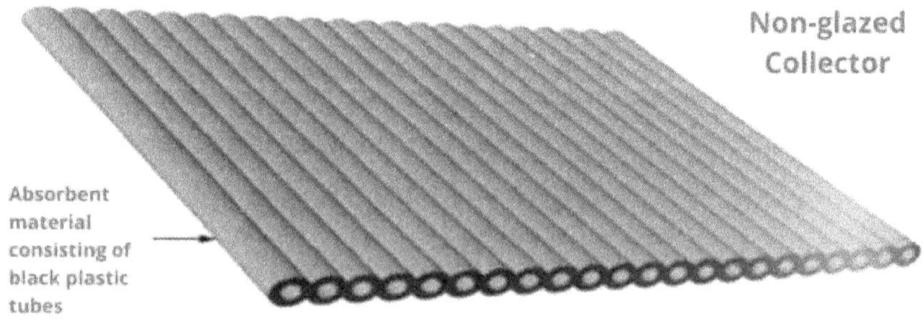


For the simplicity of the structure have a lower price than the other collectors, but the yield is lower, for this reason its use is recommended for purely summer use for swimming pools, hotels, bathing establishments.


## With vacuum tubes

They consist of a row of vacuum glass tubes, each containing an absorber (usually a copper plate) that absorbs solar energy and transfers it to a fluid that carries heat. Thanks to the insulating properties of the vacuum space, the heat losses by conduction by the absorber are very low and temperatures of about 100°C above the ambient temperature can be reached.

Vacuum panels are more complex and expensive to make than flat panels and are less resistant than flat panels, but at the same time they have a high efficiency thanks to the low energy losses obtained by using vacuum ducts. This technology is particularly suitable for use in locations with low insolation or for applications where high temperatures are required (such as space heating). They have an excellent performance all year round and are suitable for installation even in very harsh climatic conditions (low ambient temperatures, frost).

## Concentration solar collectors

They are concave collectors designed to optimize the concentration of solar energy in a specific point (fire). They are only effective in direct sunlight as they must follow the movement of the sun.

This type of collector, being able to reach high temperatures (400-600°), are used for solar generators or electro-solar power plants.

Systems with solar concentration collectors, called "solar thermodynamic" They are not pure solar thermal systems and, in any case, they can be used in large installations that are not suitable for medium to small civil and industrial contexts.

## Glazed solar panels with hot air

Air panels are solar thermal panels normally of the glazed type that circulate air instead of water inside them. Here, instead of tubes that allow the circulation of the vector fluid, there is air circulating between the glass and the absorber and between the absorber and the bottom of the panel.

The hot air then transfers its heat to the domestic water, producing quantities of hot water directly proportional to the surface of the panel.

These panels are designed in such a way that the air stays in the solar panel as long as possible because it exchanges heat more hardly than water. There may be an integrated storage tank, particularly suitable for locations with a very harsh climate. A particular type of solar air panels are the cladding panels, used for the normal finishing of the external curtain walls of both civil and industrial buildings: these are not glazed but consist of an external surface with a metal box that acts as an exchanger and heats the air passing inside the panel, which can be introduced into the rooms with a special suction system.

## Panels with integrated tank

In panels with an integrated tank, the heat absorber and storage tank are contained in a single block and the solar energy comes directly to heat the stored water. The tank, therefore, is physically covered by the absorber and, in general, inside it there is a resistance that can heat the water in case of prolonged absence of the Sun. As a result of the principle that hot water tends to rise and cold water to fall, a so-called convective motion is created inside the tank that distributes the heat captured to the entire mass of water. The panels with integrated tank are generally recommended for those areas that are not too cold and that do not have too long and harsh winters. In fact, due to their structure, as temperatures drop, the water inside the panel risks freezing and seriously damaging the system. Easy to transport and to assemble, these solar panels are relatively low cost and are perfect for thermal uses for a summer holiday home. However, compact systems suitable for all weather conditions are also available on the market.

# CHAPTER 3: Solar Thermal System

## Natural circulation (radiator)

In this system the thermovector fluid circulating inside the collector heats up in the panel in contact with the plate subjected to solar radiation, decreases its density and therefore its weight and tends to rise (exploiting the principle of gravity and the principle of convention) in a storage tank (boiler) which must be placed higher than the panel, making room for the coldest fluid inside the collector. Such a device, in other words, self-regulates itself by naturally optimising the fluid circulation for efficient operation. It is called natural circulation precisely because there are no mechanisms for transferring the liquid to the boiler, where the heat exchange with the domestic hot water that is distributed to domestic users takes place. Once the heat is released, the heat-convector liquid falls in the lowest point of the solar panel circuit.

In this way, reverse circulation is not possible as the heat remains higher and higher. This is the physical principle used in natural push radiators.

The flat collector is connected in a closed circuit with a thermally insulated tank for hot water storage. In direct exchange systems, the water exchanged is the same water that is heated in the collectors and then rises by radiator to the storage tank from which it will be taken for use.

In indirect exchange systems, a fluid (glycol and demineralised water) is heated in the solar panels and always "by radiator" circulates in an exchanger located inside the tank where the hot water is stored.

The typical application of natural circulation is the production of hot water for domestic use. A system for the production of Sanitary Hot Water with natural circulation per housing unit has on average an area of 2-5m2 of collectors and a tank of 100-200 litres.

Advantages: it is a simple, reliable and easy to install system, you only need the connection to the cold-water inlet and the hot water outlet pipe. There are no pumps, nor electronic control systems and therefore there are no connections and electrical consumption. Maintenance is reduced to a minimum. This type of system is economical.

Disadvantages: system characterized by a high thermal dispersion that reduces its efficiency when exposed to low temperatures.

Therefore, it is recommended for installation in hot countries or for activities or small summer homes. In addition, the external boiler has limited capacity, is a weight on the roof and can affect its structural seal and has a considerable aesthetic impact.

# Forced Circulation

The components of such a system are:

- solar collector

- storage tank/exchangers

- differential thermostat or control unit

- temperature probes

- circulation pump

- expansion vessel

- heat exchanger

- valves

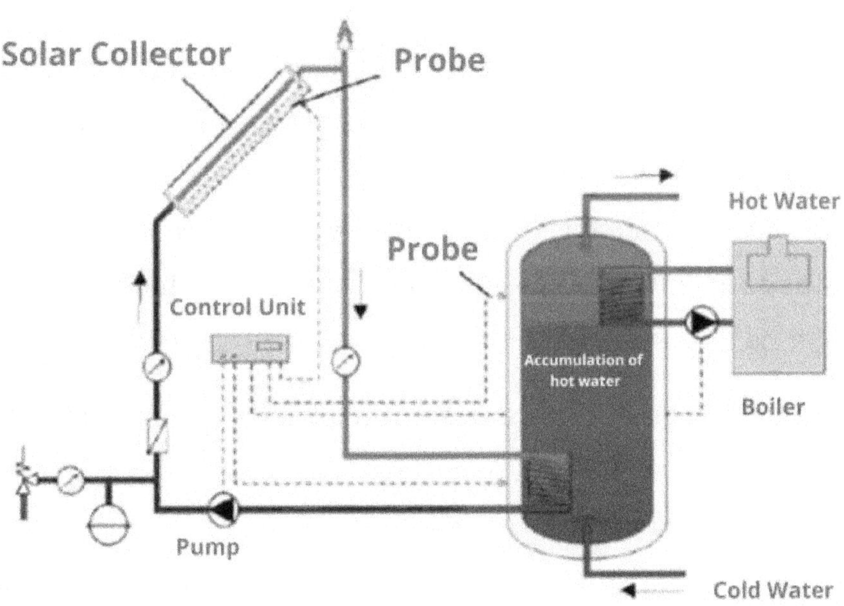

In this type of system, the tank is mounted separately from the solar panel and connected to it via a circuit. Inside the circuit there is water or an antifreeze thermovector fluid (propylene glycol) which is pushed by a pump (called circulator) set in motion by an electronic control unit able to detect the water temperature by means of special probes and which starts the pump when the temperature inside the collector is higher than the temperature set in the storage tank. In this way you have the guarantee that the system works only when it is able to really supply thermal energy and above all it avoids that through the collectors, in the absence of sunshine, the stored thermal energy is dispersed. The fluid transfers the heat from the solar panels to the water in the storage tank and transferred to the domestic hot water by means of a heat exchanger. Then there are two separate hydraulic circuits in the tank: the primary circuit of the panel, in which the fluid heated by the sun circulates and the secondary circuit in which domestic hot water circulates and which is connected to the house hydraulic system. A combined solar heating system uses a closed system, while a combined solar pool water heating system can use an open as well as a closed system.

In winter, when the solar system is unable to heat the water due to poor sunshine, the part of the tank that contains hot water that is readily available can be heated by a heat exchanger connected to a boiler. The auxiliary heating is controlled by a thermostat when the temperature of the water in the tank in the part of the tank that is ready to be used falls below the desired nominal temperature The storage tanks generally have a double coil: the one positioned at the bottom is connected to the solar circuit (inside which circulates a mixture of water with antifreeze), while the one positioned at the top is connected to the boiler. If the solar system is not able to bring the water to the desired temperature (about 40 °C), the heat supplied by the boiler through the coil (auxiliary system) guarantees the user the right integration.

In this type of system, the circuit is more complex and includes an expansion tank, temperature control and other components.

The optimal size of the storage tank allows to best meet the needs of the family and depends on climatic conditions, the type of energy demand and economic conditions. If both technical and economic aspects are taken into account, the range of optimal values is generally between 50 and 100 lt per square meter of collection area. In addition to the dimensioning, the insulation of the boiler is an important factor in the proper functioning of the system because, by reducing the energy dispersed, the energy available to the user increases. The insulation of the boilers is particularly important if they are external as it happens in natural circulation systems. A forced circulation home system per housing unit has on average an area of 3-6 m2 and a tank of 150-400 litres.

Advantages

Higher thermal efficiency compared to natural circulation because the storage tank is placed inside the house or underground, and therefore is not subject to heat loss at night or in cold weather conditions, despite the energy costs due to the use of the pump

This system can integrate environmental heating, resulting in savings on the energy bill.

Advantageous for the production of high quantities of hot water and therefore for hotel activities, sports centres, retirement homes, etc.

The disadvantages are related to the higher complexity, higher cost and need for maintenance compared to the natural circulation system.

## Emptying Circulation

The emptying circulation is similar to forced circulation, with the difference that the system fills up only when it is possible or necessary, i.e. when there is sunshine or when the tank has not reached the desired temperature. In other cases, the system remains at rest. The pump then empties it if there is no light or if the desired storage temperature has been reached.

# CHAPTER 4: What a Photovoltaic panel looks like

The name "photovoltaic" itself allows you to understand the meaning and operation of these solar panels. This word is composed of the terms "photo" (light) and "voltaic" which comes from the inventor of the electric battery, Alessandro Volta.

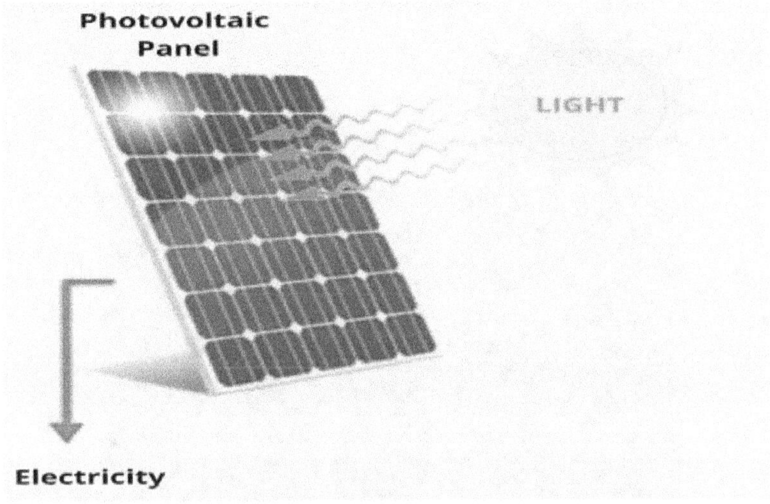


So, they should not be confused with solar panels that produce domestic water. Photovoltaic panels are a type of solar panels dedicated exclusively to the production of electricity.

A solar panel is composed of smaller units, called photovoltaic cells. Each photovoltaic cell (solar cell) transforms the sun's rays directly into electricity through the photovoltaic effect.

**Photovoltaic Cell**

The photovoltaic cells are connected together in a photovoltaic module (PV module).

## Photovoltaic Module

A photovoltaic solar panel consists of several photovoltaic modules connected in series. For this reason, when exposed to sunlight, the photovoltaic panel produces electricity.

**Photovoltaic Panel**

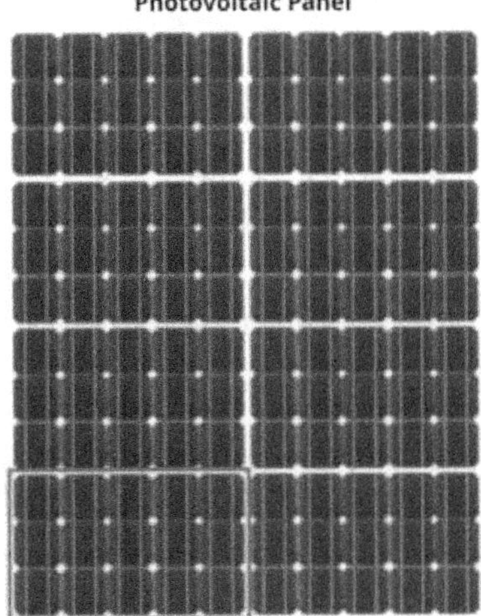

Photovoltaic Module

A photovoltaic solar system generally consists of several solar panels. It is made in a place directly exposed to the sun and without obstacles. For example, on the roof of a house, a terrace, a canopy or on the ground.

This type of photovoltaic panel involves wiring each cell, which on the surface has a grid made of conducting material that channels electrons. Each cell is connected to the others with metal strips, thus creating real circuits in series and in parallel. The silicon must be pure, so it is purified with silicon oxide (chemically $SiO_2$ silica).

In the back is placed a support surface, usually in material this time insulating and which usually has a low thermal expansion, then tempered glass, tedlar (a polymer). This surface is placed on a thin layer of vinyl acetate (EVA), a matrix consisting of preconnected modules (with metal tapes), another state of acetate and finally a transparent material, usually tempered glass, which acts as mechanical protection at the front of the cells.

All these layers are bonded by a die-casting process at 145°C in a laminator for 10 minutes, which transforms the acetate into an adhesive. The electrical terminations are insulated with a watertight terminal block, fixed to the back support. The whole is fixed with an aluminium frame, useful for attaching the panel to the supports.

The structure is then summarized as follows:

- Front glass (4 mm usually)
- EVA (ethylene vinyl acetate)
- Mono or polycrystalline photovoltaic cells
- EVA
- Bottom: Glass or tedlar

## Difference between photovoltaic panel and solar thermal panel

Apparently, the photovoltaic panel looks like a classic solar thermal panel, yet they are two different technologies. With solar thermal, photovoltaics has in common only the fact that it takes energy mainly from solar radiation, for the rest the purpose and operation is completely different, as well as cost and duration. The difference between a solar panel and a photovoltaic panel is not always clear to everyone: we will try to simplify.

The photovoltaic panel is made from silicon, a material made to absorb solar energy and thanks to other materials generates electricity. The solar panel, on the other hand, is a collector that works by heating fluid substances (usually water), then stabilizing them at a present temperature in the user. The solar thermal panel has the essential purpose of producing hot water, for heating and for sanitary purposes, such as personal hygiene and washing dishes, while photovoltaic is used to produce electricity, i.e. light.

The yield between the two types of appliance is quite different: in fact, if solar thermal panels can convert the energy of the sun into hot water with an efficiency of 80%, photovoltaic panels convert it into electricity only at 15%, at most.

## Main Applications

Photovoltaic panels allow you to produce electricity from the sun, so they can be used to produce energy in areas that are not served by the territorial electricity grid (isolated plants), or even used in covered areas in order to reduce your electricity bill, with even less environmental impact (grid-connected plants).

**Isolated systems**: Photovoltaic systems operate on the basis of the photovoltaic module, consisting of several cells, which transforms the energy contained into electricity. This can be stored in batteries, in order to make it always available for this we also speak of autonomous systems. These are used, for example, in the mountains, in agricultural areas without a grid, etc. These systems must always be dimensioned according to the area, the use, the type of user, the necessary load.

**Grid-connected systems**: The energy stored in photovoltaic modules can also be used immediately by the user, even without the need for batteries. This is the case with grid-connected systems. The peculiarity of these systems is that they have an interchange regime with the local electricity grid, i.e. when the sun is shining, the user consumes the electricity produced by his panel, while when there is not enough or not enough, the electricity grid will provide him with the necessary energy. If the plant produces more energy than is required, it can be fed (sold) into the grid: we speak of the transfer of surpluses.

# CHAPTER 5: Photovoltaic Systems

There are small, medium and large photovoltaic systems.

Small photovoltaic systems are generally installed on the roofs or terraces of houses to produce electricity and transfer it to the national grid in exchange for compensation on the bill (grid connect systems) or for self-consumption of electricity (standalone systems). In general, the panels have a low nominal power rating (up to 20 kW) and cover an area of a few tens of square meters.

Medium-sized photovoltaic systems cover an area of many tens of square meters and have a higher power than domestic photovoltaic systems. The nominal power is between 20 kW and 50 kW. They are generally grid connect systems installed by companies as a secondary activity.

Large photovoltaic systems occupy several hundred square metres and reach a nominal output of more than 50 kW. They are also called solar farms. These systems are operated by companies in the energy sector that have energy production as a primary activity. The electricity produced by the plant is sold to the national grid.

# Photovoltaic cell structure

The element at the base of photovoltaic technology is the cell which is made up of a semiconductor material, silicon, of extremely reduced thickness (0.3 mm), which is treated by means of a "doping" operation which consists of treating the silicon with phosphorus and boron atoms in order to obtain stable electrical currents inside the cell.

- Realization of the metallic electrical contacts: to the silicon layer are applied by silk-screen printing system of the metallic electrical contacts (in silver or aluminium) which consist of a continuous surface on the back side and a grid on the front side of the cell. Their function is to capture as much electrical flow as possible and convey it outside.

- Anti-reflective coating consisting of the deposition of a thin layer of titanium oxide to minimize the reflected solar radiation component.

- Texturizing: the surface is not flat, but shaped into tiny pyramids in order to increase the surface area for the capture and promote reciprocal reflections.

- The most important parameter of the cell is its efficiency "$\eta$" which represents the ratio between the maximum power Pmax [Wp] obtained from the cell and the total power of the incident radiation on the front surface.

- The level of efficiency decreases as the temperature of the cells increases, as the temperature hinders the passage of electrons into the semiconductor. ($\eta=$ **P cell/Pmax**)

# Photovoltaic cell type

Currently on the market photovoltaic cells have different sizes depending on their type.

**Monocrystalline silicon cells**: they have a higher degree of purity of the material and guarantee the best performance in terms of efficiency having the highest efficiency equal to 15%. They are uniform dark blue in colour and have a circular or octagonal shape, from 8 to 12 cm in diameter and 0.2 - 0.3 mm thick.

**Polycrystalline silicon cells**: they have a lower purity which means lower efficiency, i.e. their efficiency is between 11 and 14%. They have an intense blue iridescent colour due to their polycrystalline structure. They have a square or octagonal shape and a thickness similar to the previous type.

**Amorphous silicon**: this is the deposition of a very thin layer of crystalline silicon (1-2 microns) on surfaces of other materials, such as glass or plastic supports. In this case it is inappropriate to speak of cells, as surfaces can be covered in a continuous manner.

The efficiency of this technology is significantly lower, in the order of 5- 6.8% and is subject to a consistent decay (-30%) of its performance in the first month of life (Stabler-Wronsky effect), which requires an oversizing of the installed surface, so as to allow the production of electricity during the operation phase planned during the project.

## From cell to PV module

The photovoltaic modules are made up of several overlapping layers:

1.      tempered glass sheet of variable thickness which has a double function: to ensure good thermal transmittance (> 90%) and mechanical resistance, considering the fact that photovoltaic cells are very fragile and break easily;

2.      first transparent sealant sheet in EVA (ethylene vinyl acetate) which has the function of ensuring the seal to external agents and a good dielectric insulation;

3.      photovoltaic cells;

4.      second EVA sealing sheet for back insulation;

5.      Rear closure which can be either in glass (see modules produced by Schuco International) with the function of promoting heat exchange and allowing partial transparency of

the module, or in polyvinyl fluoride (PVF) known commercially as tedlar® which is used in sheets in the assembly of photovoltaic modules due to its special anti-moisture characteristics.

The sandwich is placed in a lamination oven where, by heating to about 150°, the components are sealed, the EVA becomes transparent and the air and steam contained between the interstices are eliminated from the inside of the stratification in order to avoid possible corrosion processes. Once the laminate has been produced, the module is completed with aluminium frames, even though recent projects have been aimed at solutions without frames, which are lighter and more preferred in the architectural field. In the back of the photovoltaic module is connected the junction box for the electrical connections required for installation.

## Photovoltaic generator

**Photovoltaic cell**: basic element of the photovoltaic generator, it consists of semiconductor material properly treated by "doping", which converts solar radiation into electricity.

**Photovoltaic module**: a set of photovoltaic cells connected together in series or in parallel, so as to obtain voltage and current values suitable for common uses. In the module the cells are protected from atmospheric agents by a glass on the front side and by insulating and plastic materials on the back side.

**Photovoltaic panel**: a set of several modules, connected in series or in parallel, in a rigid structure.

**String**: set of modules or panels electrically connected in series to obtain the working voltage of the photovoltaic field.

**Photovoltaic generator**: electric generator consisting of one or more photovoltaic modules, panels, or strings.

# CHAPTER 6: How PV Systems Work

Today everyone knows what photovoltaics is, everyone knows that it is a technology that allows you to produce clean energy using sunlight. Everyone knows that it is a renewable source that allows to reduce polluting emissions into the atmosphere. Many people now know that, as a source of clean energy, it will be the future of a new energy model that will replace the exhausted fossil fuels. You know, in short, "what is" photovoltaics, but not everyone knows how it works.

In this chapter we see how a photovoltaic system works and how it manages to produce energy using the energy of the sun. We see which are the main components and which are the factors that can compromise its performance. We will see, then, how to size the right system starting from the electricity bill, from the actual consumption and how to mitigate as much as possible the losses due to shading and other inefficiencies.

Photovoltaic panels, consisting of the union of several photovoltaic cells, convert the energy of photons into electricity. The process that creates this "energy" is called the photovoltaic effect, the mechanism that, starting from sunlight, induces the "stimulation" of the electrons present in the silicon of which each solar cell is composed.

Simplifying to the maximum: when a photon hits the surface of the photovoltaic cell, its energy is transferred to the electrons present on the silicon cell. These electrons are "excited" and begin to flow in the circuit, producing electric current. A solar panel produces energy in Direct Current (DC).

It will then be the task of the inverter to convert it into Alternating Current to transport it and use it in our distribution networks. In fact, domestic and industrial buildings are designed for the transport and use of alternating current.

## The components of a photovoltaic system

As many people know, every photovoltaic system consists of at least two basic components:

- photovoltaic modules, consisting of photovoltaic cells that transform sunlight into electricity,
- one or more inverters, devices that convert direct current into alternating current. Modern inverters integrate electronic systems for "intelligent" energy management and conversion optimisation. They can also integrate temporary electricity storage systems: AGM, Lithium or other batteries.

Here is the "basic scheme" of a simple photovoltaic system, consisting of 18 panels connected in a single string to a single DC/AC inverter.

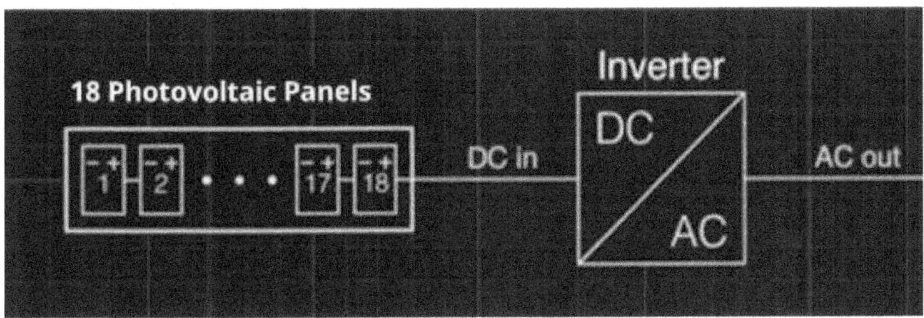


## Factors affecting the efficiency of photovoltaics

Not everyone knows that the conversion efficiency of every photovoltaic system is not 100%. That is: the panels, the solar cells, which are hit by the sun's rays, do not transform all the energy received into electricity. They can only convert part of it: this is the conversion efficiency. The best modules have a conversion efficiency of around 20-22%. This means that only one fifth of the solar energy that hits the panels is actually converted into electricity. Some "experimental" modules can achieve conversion efficiencies of over 30%, but for these the production costs are still too high.

In addition to this "physiological" factor, many others determine the actual efficiency of each system. These are both "losses" due to environmental factors and inefficiencies due to various electrical dispersions (cables, equipment, transport).

Typically, the factors that determine the yield of a photovoltaic system are:

## The temperature

The efficiency of photovoltaic modules varies according to the operating temperature: the higher the operating temperature, the less efficient the panels are. The overheating of the cells has a negative impact on the efficiency of the modules and the efficiency of the entire system.

## Dirt

The materials that can accumulate on the surface of the panels (earth, sand, pollution, bird droppings, leaves, resins, etc...) have a negative impact on the full reception of sunlight and hinder the efficiency of the PV system. In the long run they could also compromise the economic return expected from the investment plan. The yield losses due to this type of "inefficiency" can be very variable and depend a lot on the environmental conditions and the frequency of cleaning of the panels. Cleaning is, in this case, not just an "aesthetic" element, but a "functional" one.

## Shading

They can be "passenger" (at certain times of the day) and can derive from the surrounding presence of trees, other buildings or even chimneys on the roof itself. These are "calculable" inefficiencies. They have a high index of variability, instead, other passenger shading caused by clouds and the surrounding environment. However, there are technologies able to minimize the incidence of shading on the performance of the photovoltaic system. We see them later in this guide.

## Wiring and Connectors

Even the use of cables and connectors cause small yield losses. In this case, these are electrical losses that only have a minimal impact on the overall yield of the system.

## Mismatch

We could translate the term Mismatch as "mismatch" or, better, as "irregularity". What does that mean? It means that not all panels of the same brand, of the same power and of the same model, produce perfectly homogeneously. Between similar panels, subjected to the same operating conditions, there are always minimal variations in efficiency. These are minimal "factory" variations that give the panels slightly different electrical characteristics. This "mismatch" can also be one of the factors to take into account when estimating the efficiency losses of a system.

## Inverter efficiency

The process of conversion from direct current to alternating current by means of an inverter normally has an efficiency of around 96-97%. Inverters typically have an optimal conversion efficiency when the "input" DC power is high, but always below the declared rated power.

## Senility

Photovoltaic cells, which last from 20 to 25 years, do not produce homogeneously throughout their lifetime: they have a drop in efficiency which is estimated at 0.5% per year. At the end of its life, a PV system can therefore have an efficiency of about 10-12 percent lower than it had at the beginning. This depends on a "physiological" degradation of materials and components and must be taken into account from the beginning in the depreciation plan of the system.

|  | Typical Value |
|---|---|
| Temperature | -0.5% each degree centigrade |
| Inverter Efficiency | 96,5% |
| Mismatch | 98% |
| Wiring and Connections | 98% |
| Dirt | 95% (high index of variability) |
| Aging modules | -0.5% per year |
| Shading | Very variable depending on the context |

# CHAPTER 7: Efficiency and Size of a Photovoltaic Plant

## How the efficiency of a photovoltaic system is calculated

The above factors are combined into an index, to be precise: a coefficient, which serves to represent what is called the "System Derate Factor" that is: the efficiency reduction coefficient of a photovoltaic system. Some factors are (more or less) "fixed" and calculable, others are extremely variable and depend on where the system is installed.

Some simulators, including the PVWatts Calculator of the National Laboratory of the U.S. Department of Energy, consider for this index a default value of 86%. This, however, remains only an indicative value that can be extremely variable depending on the situations in which the PV system is installed.

As already mentioned, the efficiency index of a photovoltaic panel indicates how much of the solar energy that arrives on the module is actually converted into electrical current. To do this measurement, some "standard conditions" are generally considered: the so-called STC, Standard Test Conditions. These include, in the laboratory, an operating temperature of 25°C and a solar radiation of 1,000 Watt/m$^2$.

Here is the formula used to estimate the efficiency of a photovoltaic system:

> Overall Efficiency Photovoltaic Plant = (PV Module Efficiency) X (System Derate Factor)

Obviously, the production estimates of a plant are made with special software able to calculate and consider all the variables involved, including "shading calculations".

## How to size a Photovoltaic Plant starting from the Electricity Bill

As usual, we start with "needs". What is the first tool we have to understand how much electricity we need? The Electricity Bill.

Usually the bill gives us some important indications on how to correctly size a photovoltaic system: it shows us what our consumption is and, above all, in which time slots it takes place. The other element that the bill shows us is the cost: how much do we spend monthly to meet the energy needs of our home or our business?

With this important information we are already able to do our calculations to understand if and how much photovoltaic can be useful to us, if and how much it can save us, considering that the system produces a lot during the day "following" the day/night and summer/winter cycles.

The electricity bill, therefore, reveals information on electricity costs and consumption. With this information it is possible to estimate the most suitable size for our photovoltaic system that can compensate our consumption in an optimal way.

> Daily energy demand = average monthly consumption / days of the month

Assuming a monthly consumption of a small activity of 500 kWh and considering a 30-day month, we have:

> Daily energy demand = 500 (kWh/month) / 30 = 16.7 kWh/day

We will need a plant that produces, on average, 16 kWh/day. The daily production is very variable according to the seasons: in winter the plant will produce less than this average. In summer it will have to produce more. The on-site exchange mechanism will compensate (partly) for these seasonal fluctuations in production.

In addition to the daily requirements, we must consider the value of solar radiation that varies depending on where the photovoltaic is installed. The Equivalent Sun Hours are used as a reference parameter for each specific area. To identify the "equivalent hours of sunshine" there are special tables, but indicatively we can define them as the hypothetical number of hours per day in which the irradiation at 1,000 Watt/m2 would produce the energy produced on average by that area. For example: "6 equivalent hours" means that in a zone the energy received by the sun in one day is equivalent to the energy that the same zone would have received in six hours with an irradiance of 1,000 Watt/m2. It is a kind of "normalizer" to measure the production potential of a place and compare it with other places.

Graphically we can represent it in this way.

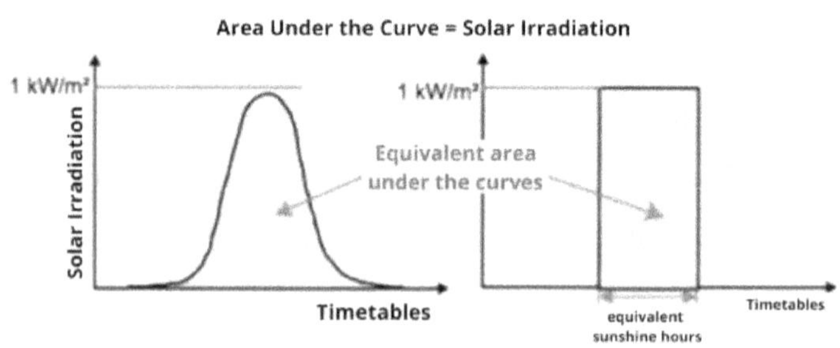

Assuming we are in an area equal to 5.2 Hours-Equivalent-Day, this is the formula to find the most suitable sizing of the plant that corresponds to our daily needs.

$$System\ Power\ PV = \frac{Daily\ Energy\ Requirement}{hours\ of\ sunshine\ daily\ equivalent} = \frac{16{,}7\frac{kWh}{day}}{5{,}2\frac{hours}{day}} =$$

$3{,}21kWp$

This would be the size of the photovoltaic system if our system had 100% efficiency. As said, it is not. For this reason, we must "correct" this sizing considering an average rate of inefficiency due to all the factors that we have already listed in this article: not only the efficiency of solar energy conversion, but also dirt, dispersion due to wiring, connectors, physiological degradation due to the age of the modules, etc.

A "standard" inefficiency rate that is typically taken into account is 0.8, but it can also be very different depending on the place and conditions of installation.

We must therefore add this correction to the formula:

> Dimensioning PV system = Theoretical power system / Inefficiency rate = 3.21 kW0.80 = 4.01 kWp

The optimal sizing of a photovoltaic system for a demand of about 16 kWh/day in an area with about 5 Hours Peak of Sun Equivalent is therefore 4 kWp. Obviously, the example does not take into account the incidence of possible shading.

## The effects of shading on the production of the photovoltaic system

Since every photovoltaic system produces electricity on the basis of the sunlight it receives, the study of shadows is a fundamental issue to calibrate every solar installation well.

The effects of the shading of a tree, for example, which perhaps only affects one panel, can be worse than you might imagine because it affects the performance of all the other modules. Contrary to what intuition might suggest, moreover, the loss of efficiency of the system is not proportional to the surface covered by the shadows. An experimental study by Stanford University has shown that by shading only one of the 36 cells of a photovoltaic panel, the output power of the module can be less than 75% of the initial power.

## How do shadows affect the flow of energy?

To understand in a simple way the incidence of shadows on photovoltaic production we can imagine the photovoltaic system as a tube crossed by a flow of running water. Just as the flow of water passing through a pipe is constant, so, with the same amount of radiation, the flow of electricity through the PV modules is constant.

Shading a solar cell is equivalent to introducing an obstacle, an obstruction, to the free flow of water in the pipe: the whole system will be affected by this obstruction. Similarly, when a solar cell, or part of the system, is "covered" by a shadow, the entire flow of electricity through the photovoltaic string is reduced (the string is the set of several panels connected in series). In this way we will have an overall decrease of the electric generation more than proportional to the shaded surface.

Graphically we could represent it in this way:

## String of Cells / PV Modules

The same mechanism of the photovoltaic cells, also occurs at the level of photovoltaic panel and string: even if only part of the photovoltaic string is reached by a shadow, even non-shaded panels, which could work at 100% of their potential, actually work at lower levels, producing less than they could.

## How to remedy the problem of shadows on photovoltaic modules

The problem of yield drops due to shading on the panels can be mitigated. It can be reduced in 3 ways:

1. using a different string configuration,
2. using the bypass diodes,
3. using the power electronics at module level (below we see what it means).

## Acting on the configuration of the photovoltaic strings

Several photovoltaic panels connected in series form a string. Several strings can be connected in parallel. If, instead of connecting all modules in series, we create more strings connected in parallel, we can reduce the impact of shading on plant production. This is one way to minimize the impact of shadows on the yield of the entire installation.

So, for example, if you install a system on the flat roof of a building surrounded by a parapet, modules that could be shaded by the parapet must be connected in a separate string. In this way, the production of the system can always be kept at an optimal level, even at times when some modules are shaded.

The mechanism is simple and with a design everything becomes clearer:

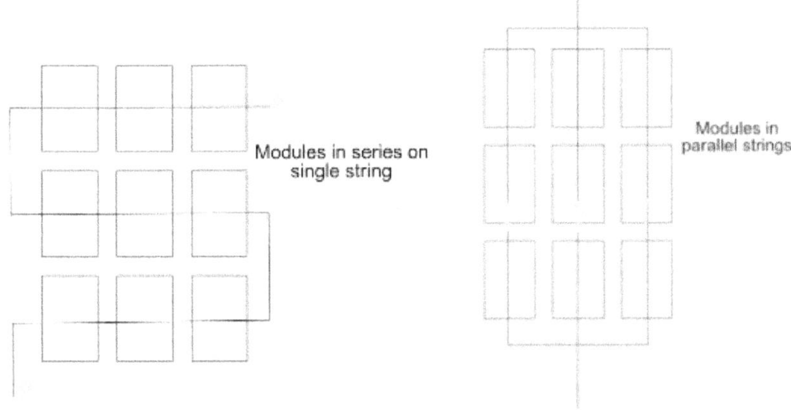


Left: system formed by all the photovoltaic panels connected in series.

Right: system with PV modules divided into parallel strings. This system makes it possible to mitigate the incidence of shading.

## Bypass Diodes

Bypass Diodes are small tools inside the photovoltaic panels that allow the electric current to "jump" ("bypass") the shaded areas of the module. In this way the electricity flows regularly inside the module even if it is obstructed in some cells. This "bypassing" occurs, however, at the cost of losing the power generated by that group of cells.

In theory, to maximize the benefit of this mechanism, the ideal would be to have a "bypass diode" for each cell of the photovoltaic module, but the operation still makes it too expensive and uneconomic. To date, for classic 60-cell photovoltaic panels, there are 3 bypass diodes positioned, as in the image below, every 20 cells.

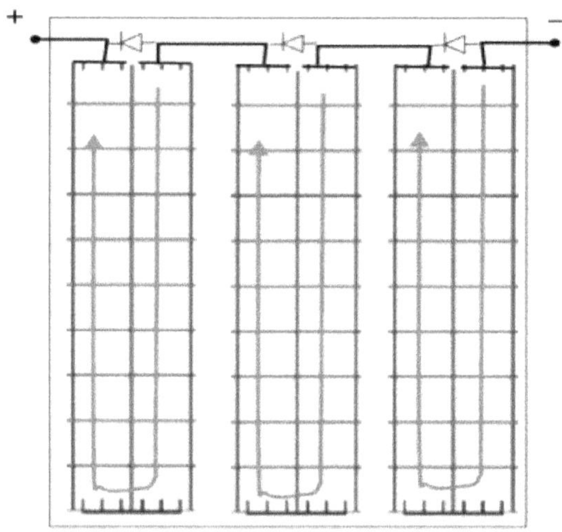

By-pass diodes on 60-cell photovoltaic module.

# CHAPTER 8: Power Electronics in Photovoltaic Systems

## Power electronics at module level

Module-level power electronics", short for MLPEs, are nothing more than devices that increase the performance of photovoltaic modules not only when they are working under optimal conditions, but also when they are shaded. In addition to being useful in this sense, they can also monitor yield at module level, indicating possible yield problems or anomalies. From a technical point of view, they are also said to be able to track and "track" the maximum power point of the photovoltaic module (MPPT - Maximum Power Point Tracking).

These are mainly two types of device:

1.      Optimizers,

2.      Microinverter

**Power optimizers** (DC side) optimize the voltage and current "output" from the modules to maintain maximum power output at module level without compromising the performance of the other panels. When a panel, for example, is reached by a shadow, the power produced decreases. The optimizer manages to reduce the module voltage to minimize the decrease in power and therefore the impact on the other modules and the production of the system itself.

**Microinverters**, on the other hand, are nothing more than many small inverters placed at the service of each module. They convert the direct current produced by each panel into alternating current and replace the use of a single inverter serving the entire string (or the entire plant). The use of microinverters allows each panel to operate autonomously. Each panel will produce "chasing" the maximum power point (MPPT) and will be completely independent of the incidence of any shading on the other modules. The malfunctioning of one module will not affect the performance of the other modules and the system itself.

On average, the performance improvement resulting from the use of microinverters or optimizers is estimated to be around 17 percent compared to the performance of systems installed with classic configurations. The incidence of microinverters and optimizers is generally very similar and in most cases is the same.

Here is the diagram showing the positioning of the Microinverters and Optimizers:

## Micro Inverter System

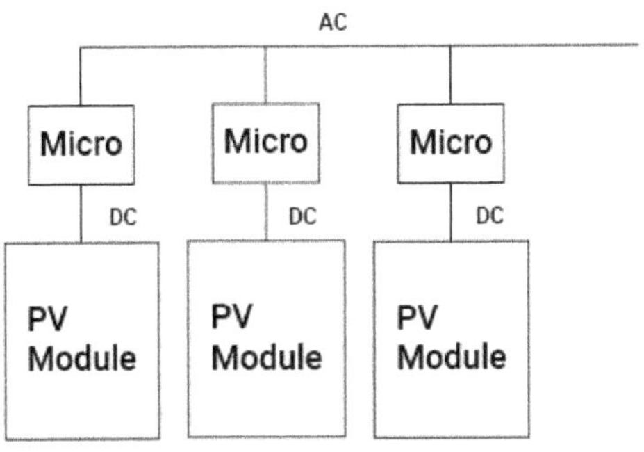


## DC Optimizer System

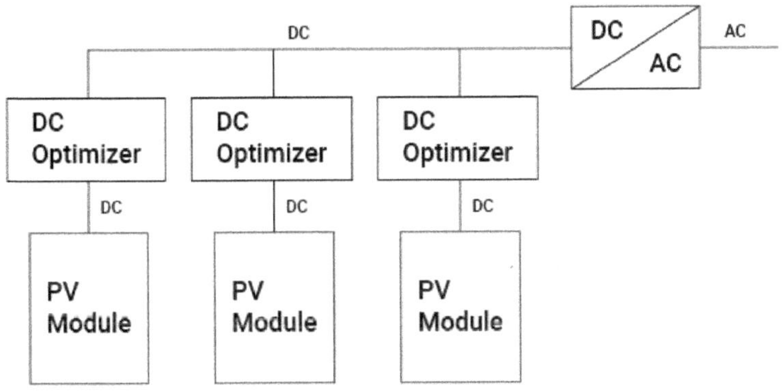

Ryan Greene
*DIY SOLAR POWER*

# CHAPTER 9: Optimizers

## Why it is important to optimize system performance at all times

Once the investment is made, every kilowatt hour not produced is a kilowatt hour lost in terms of "bill savings". Every kilowatt hour lost is a "cost in bill". The energy produced is in fact the factor that will allow you to quickly return from your photovoltaic investment. The more energy will be produced, the more the economic return on investment will be reduced. The more energy you produce, the more you will benefit from the exchange reimbursement mechanism on the spot.

For this reason, with the plant in operation, it is important to have "an eye on" these two activities:

- constantly monitoring photovoltaic production,
- increase the efficiency of the system as much as possible.

Through the use of photovoltaic optimizers, it is possible to perform both of these functions.

## What are PV optimizers?

Optimizers for photovoltaic systems are small devices, small plastic boxes, which are applied on the back of each photovoltaic panel. Each of these "little boxes" contains an electronic card that can perform some interesting functions:

- transmit, via wireless and in real time, to a control unit the production data of each single module,
- always keep the efficiency of the single panel optimal, acting "dynamically" on current and voltage in relation to the production conditions of the panel itself.

With a power optimizer, even in case of shading, dirt or other factors that hinder production, each module will be able to "always give the maximum". It will always be able to work at its "optimal working point" so as not to hinder the production of the entire string and the entire photovoltaic system.

## Let's take a closer look at how an optimizer works

The malfunction of a single panel is like a "bottleneck": the flow of electrical energy that passes and increases from module to module (the modules are in fact connected in series), can be significantly reduced if a panel does not perform its function well. The low efficiency of a single panel can be due to passenger shading, dirt or an actual malfunction or failure.

Each panel is in fact, in addition to producing direct current energy, an electrical current conductor. It receives current from the previous panel and sends it "increased" to the next panel. If the module does not work properly, the current flow produced is reduced. In this way the productivity of the entire string is compromised (a "string" is a row of photovoltaic modules connected in series).

Therefore: the malfunction of a single panel compromises the efficiency of the string in which it is located. And this, in turn, compromises the efficiency of the entire system.

For this reason, the larger the photovoltaic system is, the more it becomes necessary to optimize its production through the use of power optimizers. This useful tool will in fact be able to act in real time on individual modules with yields below their nominal power and will allow a timely and automated intervention to make them always work at their "optimal working point".

## Optimizers Utility

What are photovoltaic optimizers for

They are used to produce more energy, in all conditions. Photovoltaic optimizers increase the efficiency of the entire system by up to 25%.

As mentioned, Photovoltaic optimisers are used to ensure the highest possible productivity of the individual photovoltaic modules and the system itself at all times. In fact, each module will no longer be a simple passive "actuator", but will always follow in real time the optimal working point according to the production conditions. For most technicians, this is the same principle as the MPPT (Maximum Power Point Tracking) of photovoltaic inverters.

Not only: the optimizers also serve to monitor the production of individual panels in real time through a wireless connection to a data collection unit. Thanks to this module-level monitoring system, the individual panel responsible for the drop in production can be identified at an early stage.

As said, moreover, passenger shading or the malfunctioning of a single panel can cause the loss of production on the entire plant.

If the causes are easily identifiable on small installations, on large installations the causes of yield drops are not always easily identifiable without an adequate monitoring system at single panel level.

NOTE: the panels, in their rear connection box, can be provided with "diodes" that allow to jump any shades generated by shadows or damage of single small files of cells in the panel with a minimum loss of energy.

The panels of 60 cells are divided, inside them, in 3 mini-strings of 20 cells and each mini-string is equipped with a by-pass diode (in all 3 diodes per panel) that I estimate for the capacity of 20A (Ampere).

I think that a "PASSIVE" device of this type, which costs practically nothing if installed, is more than enough for the purpose.

# CHAPTER 10: How to Size a Home Photovoltaic Plant Not Connected to the Electricity Grid

## Theorical Calculation

The installation of a DIY Photovoltaic System not connected to the power grid, if it is a complex system, and you are not a professional in the field, it could be difficult and it would be appropriate to turn to a specialized company, instead, following step by step my instructions, you will be able to design a photovoltaic system, and buy independently the photovoltaic panels, support structures, the Inverter, the Charge Regulator, and wiring materials: reducing the overall cost without sacrificing quality, or make a comparison with the quote that companies offer you. Once you have chosen the best solution on the market you can commission your electrician to build the photovoltaic system.

## Parameters:

1. Energy required daily in kilowatt hours (kWh)
2. Calculation of the photovoltaic power (Wp=Maximum power)

3. Battery capacity in(Ah)

4. 4Charge Regulator

5. Inverter sizing according to DC side Voltage and AC side Power

6. Section of electrical cables

7. Ratio between the Yield of the Photovoltaic Panel and the available surface area.

## Daily Energy Request

The first parameter to keep in mind when you want to build a photovoltaic system for home use is the average energy requirement in kilowatt hours (kWh) per day.

First of all, I clarify what is meant by Power and what is meant by Energy.

- Power is that value in Watts given by the formula P=V x I (W).

- The Power is the Power for time, given by the formula E= P x t (kWh). The kWh is the unit of measure of electricity consumption, which is the quantity that we usually find in the electricity bill.

A simple example:

If you have a household appliance in your home that has a power of 3 kW it means that it absorbs 3 kW in the unit of time. So, if you use it for 1 h, it will consume: 3 kW x 1h = 3kWh.

To make the calculation of consumption easier, we can add up the Powers in Watts regardless of the Supply Voltage.

To sum up: we can add up the Power of a 50-Watt bulb + a 50-Watt bulb + a 100-Watt bulb, and the result will be a total Power of 200 Watt, regardless of whether they are powered at 220 V or 12 V.

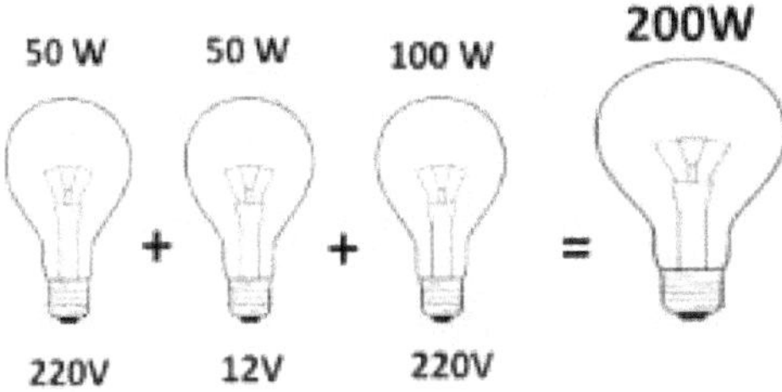


## Dimensioning example

For the dimensioning of the photovoltaic system not connected to the power grid it is good to respect a general rule: the energy produced must be greater than the energy consumed.

If you want to power with a photovoltaic system a Television, a Laptop, a Stereo, a Decoder, a PlayStation, and recharge a

Smartphone: I have to add the various Powers in Watts that I find written on the label under the device or on the instruction booklet, and then multiply them by the hours of use.

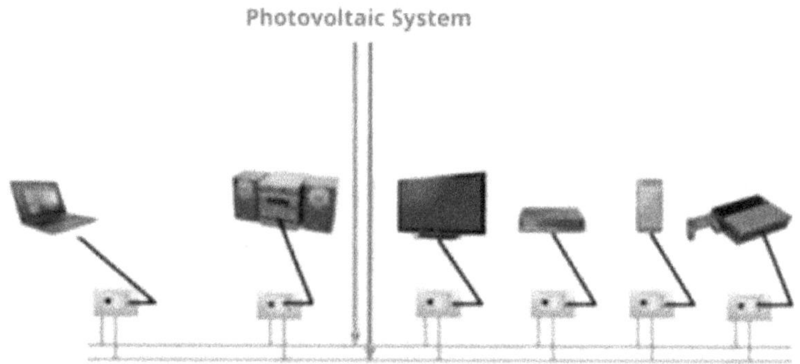


We calculate theoretical consumption:

- Television: 150-Watt x 6 hours = 900-Watt hours

- Computer: 30-Watt x 8 hours = 240-Watt hour

- 4 Smartphone: 26-Watt x 1 hour = 26-Watt hour

- Stereo: 60-Watt x 4 hours = 240-Watt hours

- Decoder: 20-Watt x 6 hours = 120-Watt hours

- PlayStation 3: 185 Watts x 4 hours = 740 Watts per hour

TOTAL: 471-Watt 2386-Watt hours (2.4 kWh)

The total contemporary power output for their operation is 471 Watts, and the total energy requirement is 2386 Watt-hours, which we approximate to 2400 Watt-hours (2.4 kWh). (See difference between kW and kWh).

# Calculation of the photovoltaic power (solar panel)

The first thing to do is to determine the period of use:

- summer only,
- only winter,
- all year round.

After that consult the PVGIS tables to check the Irradiance in kWh/m² year and get the daily equivalent hours of sunshine (hse/g).

Photovoltaic panels should face SOUTH, and be positioned to receive solar radiation for as long as possible. The inclination with respect to the ground is also of fundamental importance (to calculate the inclination according to the position see the formula), for example: with an inclination of 60° the sun's rays are better exploited in the winter period, and with 20° in the summer period; an average that is valid for the whole year is about 30°. Of course, if it were possible to vary the inclination according to the season, it would be the maximum.

The Nominal Photovoltaic Power (Wp) before system losses will be given by the daily Energy required divided by the sun hours.

PL= 2400 (Wh)/ 4.28 (hse/g= daily sun hours) = 561 Wp

As previously stated from the Theoretical Power (PL), system losses must be removed:

- Losses due to temperature deviation = 8%.
- Losses due to electrical non-uniformity between strings = 5%.
- Losses for reflection = 3%
- Direct current loss = 2%
- Loss due to battery charge and discharge=8%.
- Inverter loss = 8%
- Loss due to accumulated dirt on the modules = 2%.

## How to build a domestic photovoltaic system (theoretical calculation)

## Photovoltaic System

The photovoltaic system is an electrical system consisting of one or more solar panels that use solar radiation to produce electricity.

## Solar Cell

Solar panels are composed of many solar cells that transform light into electricity through the Photovoltaic Effect. The solar cells are composed of semiconductor material such as Silicon, which is "doped" with impure substances to give rise to Silicon type P when doped with Barium and Aluminium, and type N when polluted with Aluminium, Phosphorus and Antimony.

## Solar Cell Material

The material of the solar cells can be Amorphous Silicon, Polycrystalline, Monocrystalline, and photovoltaic cells in amorphous silicon with CIGS or CUdTe alloys. Coming soon also in Graphene.

The Amorphous Silicon module is not comparable, in an aesthetic sense, to those in Polycrystalline or Monocrystalline because the silicon is deposited evenly and in very little quantity (the thickness is a few thousandths of a millimetre) on a plastic surface or on glass, although, with the same nominal power, you must use more modules, which are also available in the traditional rigid structure or in flexible rolls.

This technology is recommended where aesthetics is preferred to energy production, which is lower than Polycrystalline or

Monocrystalline, but compensated by the cost: 30% to 40% less, and are not very sensitive to shadows.

Monocrystalline cells are dark blue and have rounded edges, these are made of monocrystalline silicon crystals all oriented in the same direction. This is why energy production is higher when the sun's rays are perpendicular, needing a smaller surface area to produce the same amount of energy as other systems, even if they are more expensive.

Polycrystalline silicon panels are less expensive than mono panels. The cells, which are bright blue in colour, are made of randomly oriented polycrystalline silicon, which means that the yield is less effective than mono, but they make better use of the sun's rays throughout the day.

Photovoltaic cells in amorphous silicon with alloys of CIGS or CUdTe are formed by a composite semiconductor material prohibited band direct, called precisely CIGS (Copper Indium Gallium (of) Selenide; (di)selenide copper indium gallium). Since the material has a high absorption power of sunlight, a much thinner film (film) is sufficient compared to other semiconductor materials. The CIGS absorber is deposited on a glass support, together with electrodes to collect the current.

## Characteristics of a solar or photovoltaic cell

In general, the characteristics are in function of these variables: intensity of solar radiation, temperature, orientation, inclination, and cell area. The physical quantities are the Irradiance measured in $W/m^2$, Voltage (V), Current (A), and the Maximum Power of the device that will be determined by the product Voltage x Current.

## From solar cell to solar panel

The solar cell is the basic component of a photovoltaic system. Each cell can produce a power from 3 to 6 Watts, little for most uses, so the cells are connected in series and soldered together through the contacts Front Blue: Negative pole, and Rear: Positive pole (negative-positive-negative-positive-etc.) and form a Photovoltaic Module. The most common Modules are made up of 36 cells that allow to obtain an output power equal to about 50 Watt, but they can also be formed by 48, 60, 72 cells that are assembled in order to have a Voltage and Current value useful to generate a Power that can reach even more than 350 Watt per module.

Several Solar or Photovoltaic Modules connected to each other form a Solar Panel.

Several Solar Panels connected in series, depending on the voltage needed to supply the electrical equipment, form a string.

## More Strings in parallel form the Solar or Photovoltaic Generator.

<u>Summary</u>

The solar cell: it is the elementary electrical component that transforms solar radiation into electrical energy.

The module: consists of several solar cells electrically connected to each other.

The panel: consists of several modules connected and positioned on the same support structure.

The string: is given by the connection in series of several panels or modules.

The solar generator: consists of several strings connected in parallel.

## Connection of Solar Panels

The Photovoltaic Generator is a set of Solar Panels that can be connected in series, parallel, or series/parallel.

If they are connected in series (called String) the Voltages (Volts) are added together, so the Total Voltage $Vu= V1+V2+V3$, etc. while the Total Current adjusts to that of the module that generates less current (in theory it remains constant).

If they are connected in Parallel it is the Current that adds up, so the $Iu=I1+I2+I3$, etc., while the Voltage remains constant.

An important note: if you put the panels in series or in parallel the Power is always the same. In series panels the Power: $P=V*I= 36*3,5= 126$ Watt.  Those in parallel $P=V*I=10,5*12=126$ Watt.

So, we will connect the panels in order to obtain the desired voltage, while with parallel circuits we will increase the current, to reach the desired power of the system.

In most photovoltaic systems we use a combination of series and parallel connections. In practice we use one or more strings of panels connected in series to increase the output voltage, and if these strings are connected together in parallel, they will increase the current and consequently the output power in Watts.

## Panels in Series and Parallel

If several strings (Solar Generator or also Photovoltaic Field) of photovoltaic modules (PFV) are put in parallel, the total current is the sum of the output current from each string (3.5 x 3= 10.5 A).

The total voltage of the photovoltaic field is that equivalent to the voltage generated by a single string (12+12= 24 V).

The theoretical total power of the photovoltaic system is equal to the sum of the powers generated by each single string.

P. Tot= (V1*I1)+(V2*I2)+(V3*I3)= 252 W

V1= (V of PVF1+ V of PVF2) =V2=V3

I1=I2=i3

P. Tot = (24*3.5) +(24*3.5) +(24*3.5) = 252 W

Or you can also do the calculation this way. As an example: if we connect in parallel three strings each composed of 2 panels of 12 V and 3.5 A, the output voltage will be 2 x 12 V = 24 V and the output current of 3 x 3.5 A per string = 10.5 A, so that the total power supplied by the photovoltaic will be 24 V x 10.5 A = 252 W, three times that supplied by a single string.

Summarizing, the number of modules that can be connected in series is quite limited for this reason to have more Power we can connect in parallel more strings, and this is done by connecting the positive pole of one string with the positive pole of the second and so on, and this also applies to the negative pole. In this case, the Power of the Photovoltaic Panels (PFV) of the first string are added to those of the other strings.

## Bypass Diode

The photovoltaic panel is made up of many solar cells put in series and the bypass diodes are used both inside the individual solar cells and between the various panels, and are used to let the current flow through each string of solar cells even in the presence of a cell or module not affected by sunlight, thus avoiding the loss of energy and prevent the reverse current from damaging the cell itself.

## How to size a home photovoltaic system not connected to the electricity grid (theoretical calculation)

The installation of a DIY Photovoltaic System not connected to the power grid, if it is a complex system, and you are not a professional in the field, it could be difficult and it would be appropriate to turn to a specialized company, instead, following step by step my instructions, you will be able to design a photovoltaic system, and buy independently the photovoltaic panels, support structures, the Inverter, the Charge Regulator, and wiring materials: reducing the overall cost without sacrificing quality, or make a comparison with the quote that companies offer you. Once you have chosen the best solution on the market you can commission your electrician to build the photovoltaic system.

As far as authorizations are concerned, also the photovoltaic system (installation, repair, replacement) is among the interventions in Free Building provided for by the Ministerial Decree of March 2, 2018, which came into force on April 22, 2018, as long as it is outside the historical centres. Therefore, for small domestic systems, if the panels are removable, they do not need any municipal authorization. If, on the other hand, the panels are fixed to the roof of the house (without changing their shape) they will need, in most cases, a prior communication to the Technical Office of the Municipality. For systems built in condominium but serving individual users (non-centralized systems) does not require any type of municipal authorization, but only that of the same condominium.

## The following parameters must be taken into account when designing a photovoltaic system:

- The Energy required daily in kilowatt hours (kWh)
- Calculation of the photovoltaic power (Wp=Maximum power)
- Battery capacity in (Ah)
- Charge Regulator
- Inverter sizing according to DC side Voltage and AC side Power
- Cross section of electrical cables
- Ratio between the Yield of the Photovoltaic Panel and the available surface.

## Daily Energy Required

The first parameter to keep in mind when you want to build a photovoltaic system for home use is the average energy requirement in kilowatt hours (kWh) per day.

First of all, I clarify what is meant by Power and what is meant by Energy.

Power is that value in Watts given by the formula $P = V \times I$ (W).

The Power is the Power for time, given by the formula $E = P \times t$ (kWh). The kWh is the unit of measure of electricity consumption, which is the quantity that we usually find in the electricity bill.

To deepen: Difference between Power and Energy, and Difference between kilowatt and kilowatt hour.

A simple example:

If you have a household appliance in your home that has a power of 3 kW it means that it absorbs 3 kW in the unit of time. So, if you use it for 1 h, it will consume: 3 kW x 1h = 3kWh.

To make the calculation of consumption easier, we can add up the Powers in Watts regardless of the Supply Voltage.

To sum up: we can add up the Power of a 50-Watt bulb + a 50-Watt bulb + a 100-Watt bulb, and the result will be a total Power of 200 Watt, regardless of whether they are powered at 220 V or 12 V.

## Example of dimensioning

For the dimensioning of the photovoltaic system not connected to the power grid it is good to respect a general rule: the energy produced must be greater than the energy consumed.

If you want to power with a photovoltaic system a Television, a Laptop, a Stereo, a Decoder, a PlayStation, and recharge a Smartphone: I have to add the various Powers in Watts that I find written on the label under the device or on the instruction booklet, and then multiply them by the hours of use.

Let's calculate the theoretical consumption:

- Television: 150-Watt x 6 hours = 900-Watt hours.
- Computer: 30-Watt x 8 hours = 240-Watt hour
- 4 Smartphone: 26-Watt x 1 hour = 26-Watt hour
- Stereo: 60-Watt x 4 hours = 240-Watt hours
- Decoder: 20-Watt x 6 hours = 120-Watt hours
- PlayStation 3: 185 Watts x 4 hours = 740 Watts per hour
- TOTAL: 471-Watt 2386-Watt hours (2.4 kWh)

The total contemporary power output for their operation is 471 Watts, and the total energy requirement is 2386 Watt-hours, which we approximate to 2400 Watt-hours (2.4 kWh). (See difference between kW and kWh)

**Calculation of Photovoltaic (Solar Panel) Power**

The first thing to do is to determine the period of use:

- summer only,

- only winter,
- all year round.

After that consult the tables or use the ENEA simulator, PVGIS to check the Irradiance in kWh/m² year and get the daily equivalent hours only (hse/g).

For example, in Lombardy, with an inclined surface of 30°, it is 4.28 kWh/m² of irradiation per day, which corresponds to an average day sun equivalent of 4.28 (hse/g).

Photovoltaic panels should be facing SOUTH, and positioned so as to receive solar radiation for as long as possible. The inclination with respect to the ground is also of fundamental importance (to calculate the inclination according to the position see the formula), for example: with an inclination of 60° the sun's rays are better exploited in the winter period, and with 20° in the summer period; an average that applies throughout the year is about 30°. Of course, if it were possible to vary the inclination according to the season, it would be the maximum.

The Nominal Photovoltaic Power (Wp) before system losses will be given by the daily Energy required divided by the sun hours.

PL= 2400 (Wh)/ 4.28 (hse/g= daily sun hours) = 561 Wp

As previously stated from the Theoretical Power (PL), system losses must be removed:

- Losses due to temperature deviation = 8%.
- Losses due to electrical non-uniformity between strings = 5%.
- Losses for reflection = 3%
- Direct current loss = 2%
- Loss due to battery charge and discharge=8%.
- Inverter loss = 8%
- Loss due to accumulated dirt on the modules = 2%.
- TOTAL LOSSES = 36%

The total efficiency will be: 100-36= 64%. So, the Photovoltaic Effective Power (PFV) will have to be increased by the loss, and will be:

- PFV= PL/η system (η system=64%) = 561/0,64= 876 Wp

The general formula is:

- PFV (Wp)=(Wh/hse/g)/0.64
- PFV= Actual PV power in Wp; Wh= watt-hours used; hse/g (irradiance)= sunny hours; System losses= 0.64.

# Automatic calculation of the photovoltaic nominal power (wp)

The Voltage and Current, continuous side, is chosen according to the Photovoltaic Power

Example:

- Up to an effective power of 200/300 Wp, 12 Volt solar panels can be used.
- From 200/300 Wp up to 1.000/1200 Wp 24 Volt solar panels can be used.
- From 1000/1200 up to 3000 Wp you can use 48 Volt solar panels.

The Solar Panels available with 12 Volt voltage are 50, 60, 70, 80, 120 Wp.

Solar Panels available with 24 Volt voltage are 35, 50, 80, 100, 150, 200 Wp.

In our example, for a Power of 876 Wp we can dimension the Voltage at 12 or 24 Volt:

- With a voltage at 12 Volt the circulating current will be 73 Amps.
- With a 24 Volt Voltage, it will be 36.5 Amps.

I=PFV/V= 750/12= 73 A

I=PFV/V= 750/24=36,5 A

For our case we could use 2 strings of 2 modules each in parallel. Each module will have a Power of 250 Watt and a voltage of 12 Volt in Monocrystalline Silicon with high efficiency. The Total Power will be equal to: 1000 Watts and a voltage of 24 Volts.

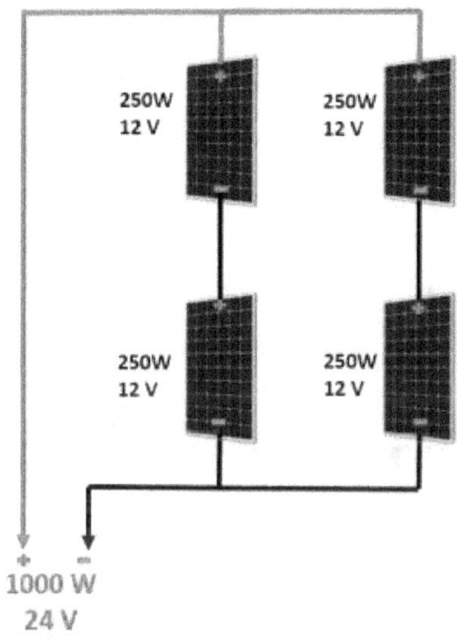


## Photovoltaic module features

Maximum Power (Pmax): 250Wp; MPP electric current: 5,95 A; MPP Voltage (Vmax): 42 V; Short Circuit Current (Isc): 6,15 A; Open Circuit Voltage (Voc): 49,40 V; Insulation Voltage: 715 V; NOCT (800 W/ m²- 20°C-AM 1,5): 47°C; Dimensions (mm) 1580 x 808 x 35; Weight 16 Kg; Cells 72 (in high efficiency monocrystalline silicon).

In most cases the Photovoltaic Modules are equipped, in addition to the junction box containing the By-Pass diodes, also the cables, and the MC4 wiring connectors, which can be male/female or Y-shaped.

## Connectors MC4

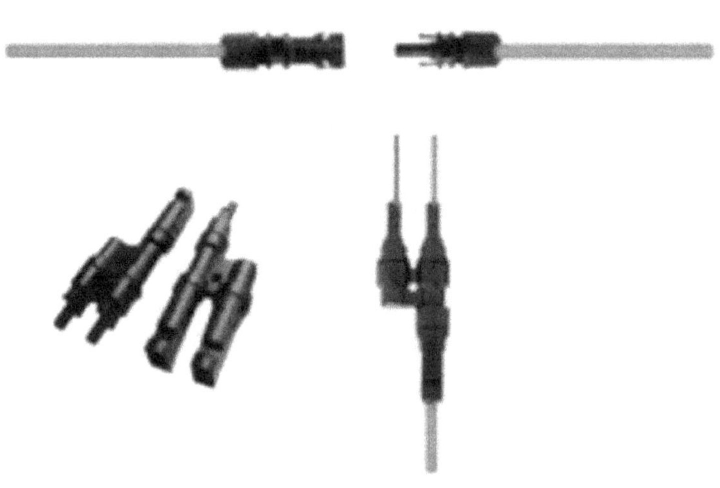

# Solar panels connected in series and parallel

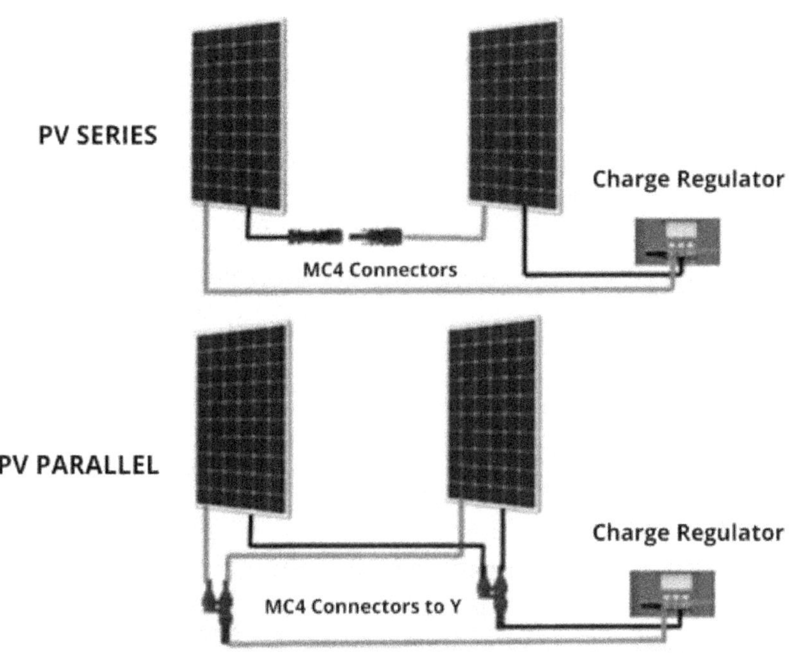


# Difference between Power and Energy

Generally speaking, in electrical systems, where there can also be energy storage such as batteries, you have to calculate how much energy has been consumed in a day in total in order to size solar panels for example, first you have to identify the energy used instantly and then multiply it by time.

I would like to make some clarifications on Electricity, Power, and Energy.

Ampere (A) is the measure of Current, while Power is produced by Current and is measured in Watts (W) , and the formula that binds them is: Power(P)= Current (A) x Voltage (V), so if the Current remains unchanged and the Voltage varies, for example: 15 Ampere and 20 Volts, produce 300 Watts of Power, while if the Current remains the same but the Voltage varies, for example 48 Volts, the Power will be 720 Watts, the difference is enormous. So, the Current (A) is used to size cables, regulators, and partly also batteries, while the Power (W) is used to understand how much "work" the Current produces with the Voltage you choose.

All this reasoning is fine when it happens instantly, but it is clear that a 50 Watt bulb kept on for one hour consumes less than the same bulb kept on for 2 hours, even if the Power is always 50 Watt, so I need to find a method with which I can do the calculations in a simple way, comparing the consumption or production as if it had happened in one hour. This ratio is the Watt hour (Wh) even if the most used is the kWh that is Energy and is equal to the Power for time (E=P x t) so the 50 Watt bulb kept on for one hour will consume 50 Wh of Energy and that kept on for 2 hours will consume 100 Wh.

POWER and ENERGY:

The electrical power (P) passing through a circuit section is given by the product of the voltage (V) and current (I) present in the section considered:

- P=V×I (Unit of measurement of power is the Watt, abbreviated as: W)
- Electricity is the product of power (P) for time (t). Unit of measurement of energy is the kilowatt hour, which is abbreviated to kWh.

**E=P×t (kWh)**

Summary

A = current flowing at this moment, is like the flow of water in a pipe that supplies the container.

Wh or kWh = total energy used, produced or stored, is like the bank account.

W = instantaneous power produced or consumed; is how much work we can do at the same time.

# Difference between Kilowatt and Kilowatt hour

Kilowatt (kW) is a unit of measurement of Power, in other words kW is by definition the amount of Energy absorbed in the unit of time. A kW, equal to 1,000 Watt, identifies the unit of Electrical Power (W=J/s) and represents the amount of Energy (Joule) in time (seconds).

Kilowatt hour (kWh) is the unit of measurement of Electricity. In fact, the Kilowatt hour is formed by the terms Kilowatt and hour, and indicates the Energy supplied in one hour of time with the power of one Kilowatt.

The two units of measurement are different because kW measures power while kWh is the unit of measurement of electricity consumption.

A simple example:

If you have a household appliance in your home that has a power of 2 kW it means that it consumes 2kW in the unit of time. So, if you use it for 1 h, it will consume: 3 kW x 1h = 3kWh.


POWER and ENERGY:

- The electrical power (P) passing through a circuit section is given by the product of the voltage (V) and current (I) present in the section considered:

- $P=V\times I$ (Unit of measurement of power is the Watt, abbreviated: W)

- Electricity (W) is the product of power (P) for time (t). Unit of measurement of energy is the kilowatt hour, which is abbreviated to kWh.

**$W=P\times t$ (kWh)**

# CHAPTER 11: Battery

## Battery capacity

If we don't want to waste the power that the photovoltaic system produces and have it available even when it doesn't come from the solar panel: the battery is indispensable, and is the most critical element of the system as it is the only part that needs maintenance. The duration is about 7/8 years and depends on the number of charge/discharge cycles and the reduced self-discharge.

In my opinion, the batteries that are used for photovoltaic systems (stationary) still more suitable today are the lead batteries, suitable to work with limited currents for long time both for charging and discharging and the price and performance ratio. Lithium batteries, which represent the latest generation, have a high charge and discharge cycle (about 6000 cycles), but on the other hand their cost is still high.

## How to calculate the battery capacity (ah)

The capacity of a battery is a measure of the amount of electrical energy it can store, and is expressed in Ampere-hour (abbreviated to Ah).

For most deep cycle batteries, it is good practice to discharge them at 50% of their nominal capacity if they are lead-acid batteries, or 80% if they are lithium batteries. (See Mathematical calculation).

## Manual calculation

To calculate the capacity of the battery pack in Ah to be used in a photovoltaic system, you must:

- Add the Watts of each device connected to the photovoltaic system and multiply it by the hours of use, and you get the Wh (e.g. the television consumes 150 Watt x 6 hours of use = 900 Wh; etc.), the result will be multiplied by the days for which it is believed that the batteries deliver energy when there is no energy from solar energy (e.g. 6 days). Therefore:  Wh total= Wh x days of use (e.g. Wh total= 900*6= 5400 Wh).
- If lead acid or lead gel batteries are used, they must be discharged at 50% of the nominal capacity, or at 80% if they are lithium batteries.
- Know the voltage of the photovoltaic system (e.g. 12, or 24 Volt)
- Efficiency: 1.1 for Gel batteries, 1.15 for AGM batteries, 1.2 for lead acid batteries.

### Capacity formula

With discharge at 50% of nominal capacity if they are lead acid or lead gel batteries.

- C (Ah)= ((Wh total * days)/ Photovoltaic Voltage))/0.5

- If we also want to take into account the efficiency of the battery, I will use the value of 1.15 which are those in AGM technology (1.1 for gel batteries, 1.15 for AGM batteries, 1.2 for lead acid) the value obtained can be defined reliable.
- C (Ah)= 1.15 *((Wh total * days)/ Photovoltaic Voltage))/0.5

With discharge at 80% of the nominal capacity if they are Lithium batteries (in this case we do not take into account the Efficiency)

- C (Ah)= ((Wh total * days)/ Photovoltaic Voltage))/0.8


## Storage capacity calculation (ah)

If the use is only for the summer period it is sufficient that the battery keeps the charge for 3 or 4 days, if we need it also in winter it is the case to double the days. For the total capacity it is sufficient to multiply the daily energy requirement in watt-hours or kWh by the number of days for which you intend to maintain the charge.

Total energy x days: 2400 x 4= 9600 Wh (summer); 2400 x 8= 19,200 Wh (winter).

(2400 Wh is the theoretical consumption we have hypothesized: TV, computer, etc.).

I will use 6 Days, which I think is good for both seasons.

Using the automatic calculator for the storage capacity I'll insert as Wh: 2400, as 24 Volt photovoltaic voltage, and 6 days. The result will be of: 750 Ah if I dump it at 80%, and 1380 Ah if I dump it at 50%.

For this type of systems, you could use lead acid or lead gel batteries (e.g. AGM) which, as I mentioned before, must be discharged at 50% of the nominal capacity, or at 80% if you use Lithium batteries.

# Charge and discharge time of a battery

## Battery Connection

The batteries can be connected in Series, Parallel, or Series/Parallel. In series you add the Voltage and the Ampere-hour (Ah), in Parallel you add the Ampere-hour and the Voltage remains unchanged, in Series/Parallel you add both the Voltage of the branch in series and the Ampere-hour of the branch in parallel.

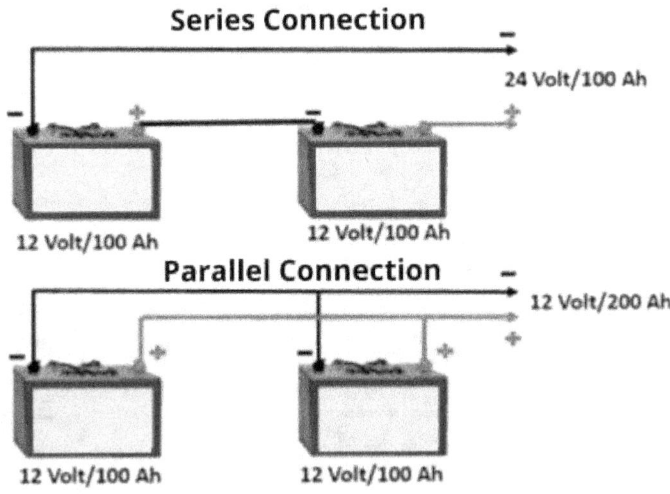



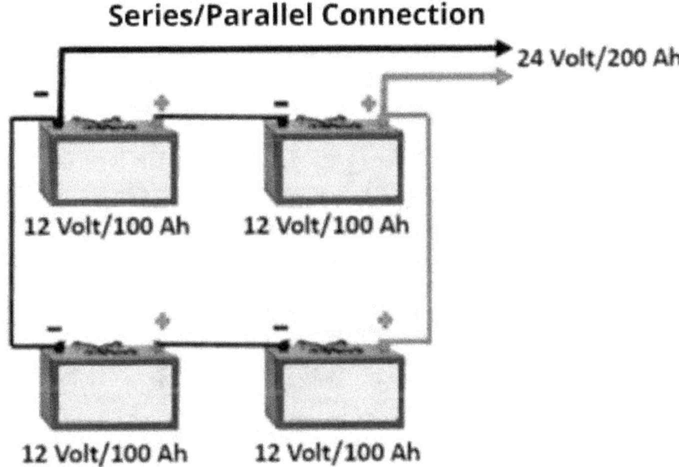



The capacities of the batteries on the market are of: 120 Ah, 12 Volt; 157 Ah, 12 Volt; 200 Ah, 12 Volt; 400 Ah, 12 Volt; 200 Ah, 24 Volt; 240 Ah, 6 Volt.

For our project we could use 4 batteries (Lithium) of 200 Ah, 24 Volts in parallel, with discharge at 80% of its nominal capacity.

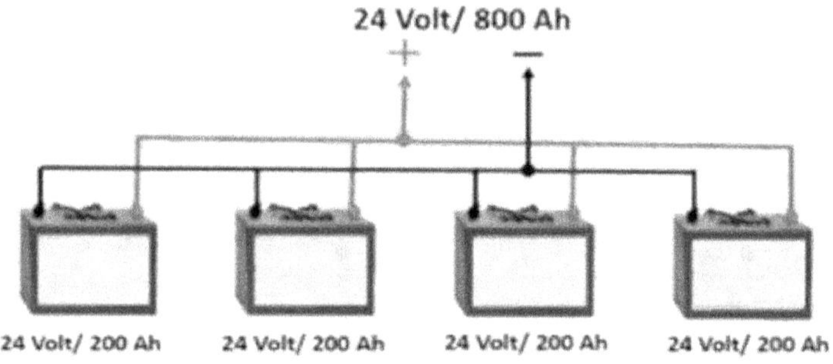


Or we could use 7 batteries of 200 Ah, 24 Volt Lead Acid (AGM), in parallel, with discharge at 50% of its nominal capacity. And that's the solution for our project.

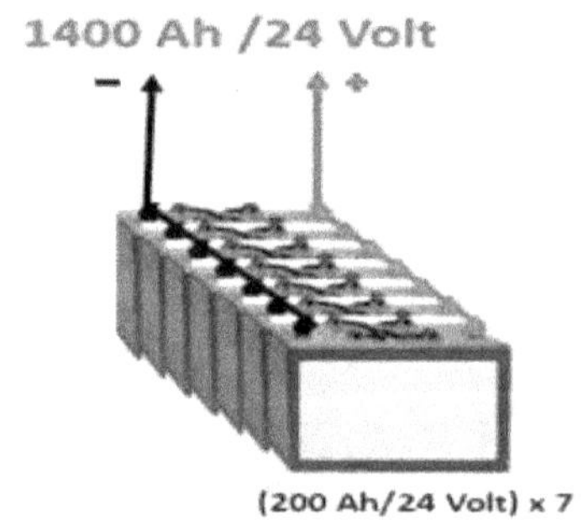


# Charge Controller

The output voltage from the solar panel is not regular and can also be high. For this reason, it is necessary a controller or charge regulator, whose main function is to adjust the Voltage adjusting it to the one needed to charge the battery, avoiding overloads. The controller is inserted between the battery and the solar panel, and will stop sending electricity to the battery once it is charged, or will exclude the load (e.g. bulbs) if the battery is in deep discharge.

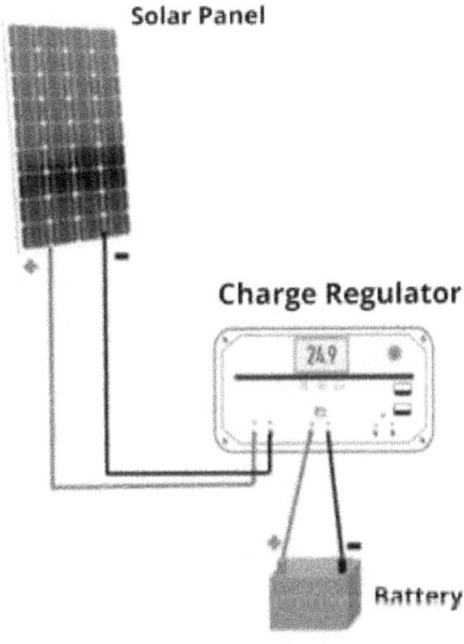


The sizing depends on the type of equipment we intend to use: PVM or MPPT.

If we use a PVM type charge regulator we have to verify the total short-circuit current of the modules or module (Isc) that you can find in the data sheet (the current Isc in the panels in series remains the same, in the panels in parallel they add up) that must always be lower than the maximum current (A) that the charge regulator can bear.

If we use an MPPT type charge regulator, we must check the maximum power (Wp) of the installed modules and that it is, at the most, that which is indicated in the data sheet, based also on the battery voltage (e.g. to an MPPT charge regulator of 30 Ampere capacity, and with a 12 Volt battery, we can, at the most, connect a photovoltaic power of 360 Watt (P= V x I= 12 x 30= 360 Watt), and 720 Watt with a 24 Volt battery. Using the formula: P=V x I).

## Choice of Controller

The choice of PWM or MPPT technology depends on the type of panels used, both on the battery bank.

PWM costs less but has limitations compared to MPPT. The most important limitation is that MPPT fully exploits the power of the photovoltaic panel using voltages higher than the battery bank, while in PWM it is not possible to connect a 24 Volt photovoltaic system with 12 Volt batteries. Therefore, in principle, it is advisable to use the PVM when the voltage of the photovoltaic panel is slightly higher than the battery bank (example: a 12 Volt panel consisting of 30 cells and 12 Volt batteries).

## PWM

In our case, having two strings in parallel and each string has 2 panels in series, we will submerge the Short Circuit Currents (Isc) of each string, indicated in the data sheet, which is 6.15 Ampere. So, the sum is: 6.15 x 2 = 12.3 A, and to rest assured we will use a higher one based on those available on the market: 6 A, 10 A, 12 A, 20 A, 30 A, 45 A, etc. 30 Ampere is the one we could use.

The photovoltaic panel is the only electric generator that when short circuited will not be damaged, but delivers the maximum current that the cells can generate.

One thing to keep in mind: never short-circuit batteries or other electrical generators!!!

## MPPT

The range of an MPPT regulator is calculated taking into account the Maximum Power (Watt) of the Photovoltaic and the voltage (V) of the batteries. Therefore, the current (Ampere) must be equal or less than or equal to the maximum MPPT flow rate stated in the specifications.

Example: in our case the Photovoltaic Power is 1000 Watts and the battery pack is 24 Volts. Using this formula, we will find the value:

- Imax= P photovoltaic (W) / Battery voltage (V) = 1000/24= 42 Ampere

For our system I will choose a 60 Ampere MPPT Regulator.

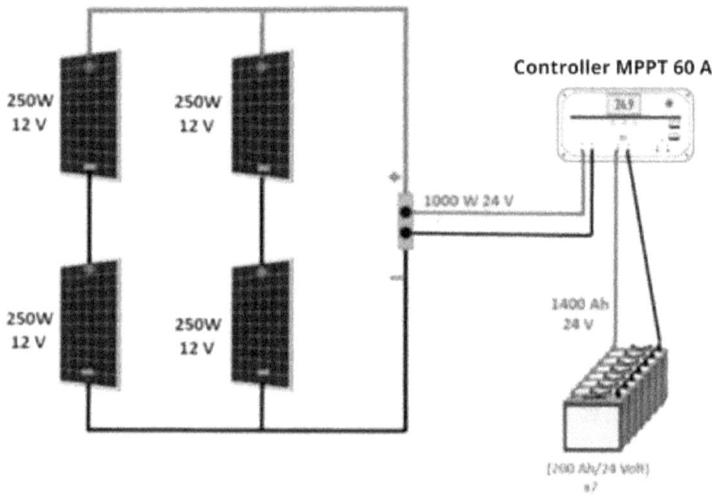

# CHAPTER 12: Inverter Sizing

If I want to use the Photovoltaic system only for lighting, I can use the 12/24 Volt bulbs that I will connect to the appropriate output of the regulator.

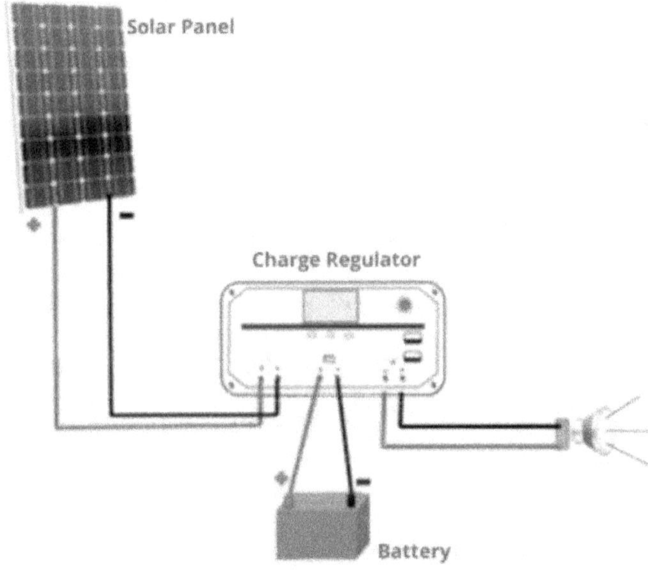


But if I want to power other equipment, I have to use an Inverter that transforms direct current into alternating current at 220 Volt, 50 Hz, which will be connected to the battery pack.

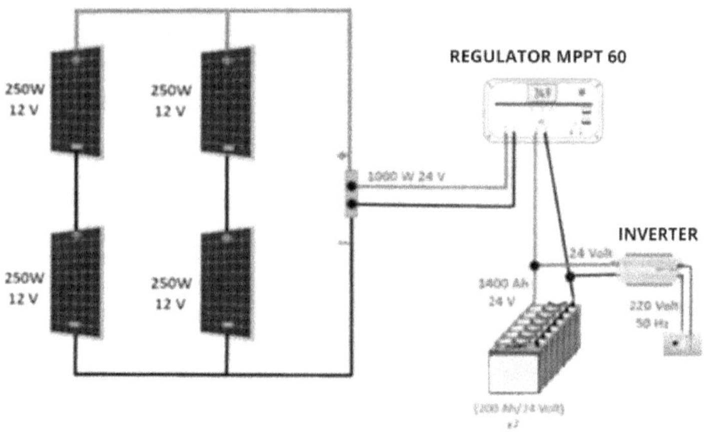


NOTE: Over 6 kW the AC output is three-phase (380 Volt).

## Inverter Selection

The dimensioning is done according to the maximum power required by the user, taking into account the ideas. For example, in our case, if all the devices work simultaneously, there would be 471 Watt without any cues.

- Television: 150 Watt
- Computer: 30 Watts
- Smartphone: 26 Watt
- Stereo: 60 Watts
- Decoder: 20 Watt
- PlayStation 3: 185 Watts

TOTAL: 471 Watt

The choice is based on the equipment available on the market, which can be square wave or sine wave. If you choose sinusoidal you can opt for a Stand-By device.

In our case we have abundant and we will use one of 1000 Watt peak, pure sine waveform output, Output voltage: 220V, Input voltage: 12/ 24/ 48 Volt, Output frequency: 50Hz, Conversion rate: 93%, with 6 types of intelligent protection against overload, high voltage, low voltage, overheating, reverse polarity, short circuit.

So, we could install an Inverter of 1000 W, 24 Volt (If there are pumps, or other devices with current inrush at power on, we need to double the rated power, in our case we are more than okay being already double).

The Inverter could also have the automatic exchange function that allows you to use the external power grid (AC (Alternating Current) as an emergency backup, so as to make maximum use of alternative energy sources. Or intervene when the power supply fails, and operate as a UPS.

**NOTE**: the electrical voltage produced at the output of the inverter is of dangerous value (230Volt) and therefore all electrical connections **MUST BE** <u>made by experienced and qualified people.</u>

# CHAPTER 13: Short Circuit and Overload Protection

When designing a photovoltaic system, it is also necessary to provide both disconnection and protection devices for all installed products, even if some already have protections such as the Inverter and the Charge Regulator inside.

The International Electrotechnical Commission (IEC) recognizes that the protection of photovoltaic systems is different from that of standard electrical installations.

In general, the following factors could be used to select fuses, e.g. for string protection, although all parameters should be taken into account with a thorough study: 1.56 for current and 1.2 for voltage. These values cover most variations due to installation.

As a rule, all PV systems that have three or more strings connected in parallel will be given protection for each string. If they have less current that they can generate is not able to damage the modules in case of failure, all this applies if the conductor is correctly dimensioned, I would put them equally on both the positive and negative wire. If the strings in parallel are from three forward one fuse in each string will protect cables and modules from overcurrent failure. For safety, in addition to the fuse, a disconnect switch is also useful, allowing you to work quietly downstream of the panels in the event of a fault.

## Fuse Specifications

If the strings in parallel are less than or equal to three, it may be sufficient that the cable is properly sized, and must be at least equal to: 1.56 x Isc (Short-circuit current), but as said before I would put a fuse in each string for safety, which must have the following parameters according to it:

- at rated current: 1.56 x Isc

at rated voltage:

- 1.20 x Voc x Ns (Ns=number of modules in series per string).

If the strings are greater than three, the fuse for each string must have the following parameters according to them:

- at rated current: 1.56 x Isc
- at rated voltage: 1.20 x Voc x Ns (Ns=number of modules in series per string).

In our case Isc= 6,15 A, Voc= 49,40 Volt. Therefore, the fuse must have the following characteristics:

- Rated current: 1.56 x 6.15= 10 Amps (approx.)

- Rated voltage: 1.20 x 49.40 x 2= 120 Volts (approx.)

Fuses can be plugged into MP4 connectors, or into fuse holders.

The Disconnector is used to "disconnect" a part of the electrical system that we want to isolate without the risk of accidental reconnection, and allow maintenance without risk for the installer, either as additional protection of the equipment from overvoltage or short circuits, in our case is 50 A.

## Photovoltaic Generator Protections

If the strings in parallel are less than or equal to three it might be enough that the cable is properly sized, but as said before I would put a fuse in each string on both the positive and negative wires, which in this case are 10 A each.

- A specific fuse for the direct current between the charge regulator and the battery, or the battery, the amperage (A) must not exceed the capacity of the regulator.
- Fuse to be installed between the output of the charge regulator (terminals where there is the bulb symbol) and the 12, 24, or 48 Volt users. The amperage (A) shall be calculated from the maximum consumption in Watts of the connected consumers which, however, shall not exceed

the maximum capacity of the regulator using the formula: $I=P/V=A$.

- A fuse or a circuit breaker (specific for direct current) on the connection starting from the battery pack, and the input of the Inverter. The flow rate (A) must be calculated according to the power in Watts of the Inverter and the supply voltage (Volts), according to Ohm's law: $P=V \times I = $ Watt, from which the circulating current $I=P/V=$ Ampere is derived (example: if we use an Inverter with a Power of 1000 Watt powered at 24 Volts, we will have the circulating current $I=P/V=1000/24= 41$ Ampere).

- A magnetothermal switch suitable for alternating current between the Inverter output (220 Volt/ 50 Hz) and the sockets of the system where the 220 Volt powered equipment (Computer, Television, Router, Washing machine, etc.) will be connected. The Ampere flow rate will be calculated based on the Inverter Power in Watts and the output voltage using the formula: $I=W/V$ (example: if the Inverter Power is 1000 Watt and 220 Volt is the output voltage we will have a current of: $I= W/V= 1000/220= 4.5$ Ampere, which will be the capacity of the magnetothermal).

# Surge arresters

The solar panels of photovoltaic systems occupy a space that is proportional to the power to be obtained, and when the occupied area becomes significant the systems are more subject to the effects of lightning strikes, especially indirect ones. To avoid damage, it would be good to install surge arresters for each polarity to earth in the place closest to the strings. The choice of the voltage of SPD (Surge Protection Device) DC side surge arresters in ground insulated systems can be calculated using this formula:

VC (SPD)= VOC STC x K

In our case the open circuit voltage (VOC) is 49.40 Volts which multiplied by K (1.20) results that the VC (SPD) is equal to: 60 Volts, and we could use OVR PV arresters.

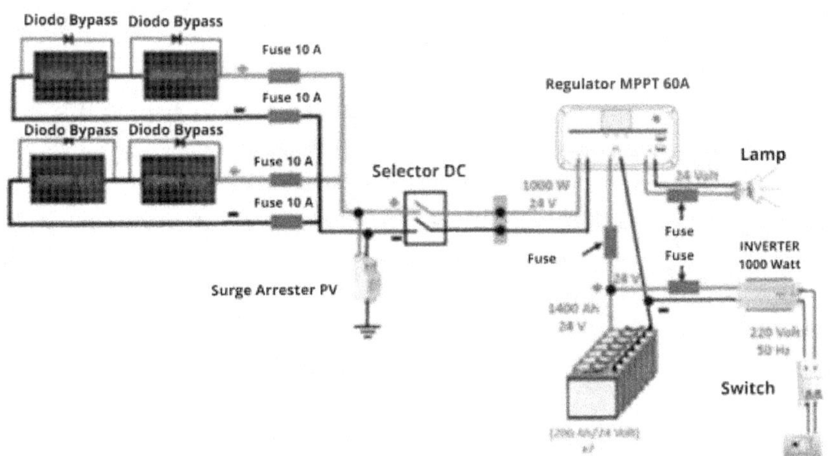

# Photovoltaic junction box

A more professional system for connecting solar panels to the voltage regulator is the photovoltaic junction box.

This device greatly simplifies the input and voltage regulator wiring.

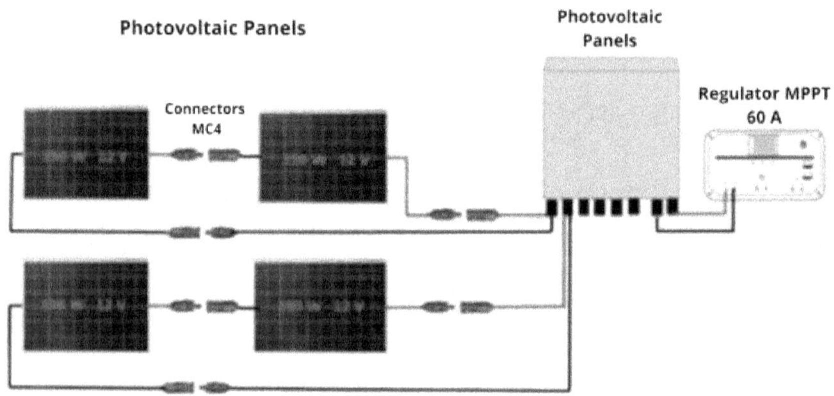


It also contains protection with a 10 Amp fuse for each string, overvoltage protection, overcurrent protection, anti-reflux diodes, anti-reflection and anti-reflection protection, safety circuit breakers, lightning/overvoltage protection (1000 Volt), and grounding.

# CHAPTER 14: Characteristics of Electric Cables

Choosing the right section is important because the current flow in the cable produces heat and if it is not well dimensioned it could overheat.

The core of the cable is made of copper and covered with an insulation sheath and must withstand temperatures ranging from -40 degrees to +120 degrees centigrade. The cables must have a nominal voltage of 1000 Volt for alternating current, and 1,500 Volt for direct current.

The minimum cross-section must not be less than 0.25 mm²/A for cables up to 50 metres long. In principle we can say that the current should not exceed 4 A / mm². So, if it absorbs 40 A, a conductor of at least 10 mm² (40/4=10) is required.

When designing a photovoltaic system, the standard requiring a minimum cable cross section of 1.5 mm² must be observed.

This in general, now let's try to understand how to reach the most suitable cable section starting from the basic concepts:

- Ohm's law, which relates three electrical quantities:
  1. Voltage (Volt),
  2. Current (Ampere),

3. Resistance (Ohm) with the following formula:

$$V=R*I$$

We can also say that fixed a current intensity I, an electric cable produces a voltage drop that is directly proportional to the resistance R. Therefore, the greater the resistance of the cable, the greater the voltage drop caused by a certain current flowing through it.

Since the voltage drop is an undesirable effect, it will be our task to try to reduce the resistance of the cable to a minimum, but it must be taken into account that the resistance of a cable increases with its length and decreases as its section increases, as can be seen from the formula:

$$R=K*L/S$$

Where:

- K is the specific resistivity of the cable, which in Copper is equal to: 0.0175 Ohm*meter;
- L is the length of the cable, expressed in meters (m);
- S is the section in mm$^2$.

While the voltage drop is calculated with the following formula:

$$cV = (K * I * Lc)/S$$

Where:

cV = voltage drop;

- K is the specific resistivity of the cable, which in Copper is equal to: 0.0175 Ohm x meter;
- I is the current passing through it,
- Lc is the total length of the cable (round trip) expressed in meters (m);
- S is the section in $mm^2$.

If we wanted to make an example and apply the formula to a copper cable of 1 $mm^2$ section, and this is crossed by a direct current, we would have a resistance of 0.0175 Ohm for each meter of length. So, assuming we have a 1-meter long cable with a section of 1 $mm^2$, we will have a resistance of 0.0175 Ohm (0.0175*1/1=0.0175).

Now let's see how to calculate the voltage drop if we had, as in our case, a 1000-Watt Inverter and a battery bank voltage of 24 Volts, and a 1-meter long cable with a section of 1 $mm^2$. First, we calculate the circulating current using the formula:

$$I = W / V = 1000/24 = 41.6 \text{ Ampere}$$

then we calculate the voltage:

$$V = I \times R = 41{,}6 \times 0{,}0175 = 0{,}728 \text{ Volt}$$

The power with the voltage drop will be: W=V x I= 0,728 x 41,6= 30,28 Watt (which is the power loss) so with a copper cable 1 meter long and with a section of 1 mm² I will have a loss of 30 Watt of power. If we were to double the section the loss would be halved (15 Watt), and so on. This parameter is important for the design of the system.

There are two solutions to decrease the voltage drop: either increase the cable cross-section, or decrease the length.

In summary, to calculate the section of a cable, knowing its total length, the current passing through it, and fixing the desired voltage drop, the following formula is applied:

$$S = K * I * Lc / cV$$

Where:

- K is the resistivity and in copper is equal to 0.0175 Ohm*meter,

- S is the cross-section of the cable in mm², 

- I is the current passing through it (in Ampere),

- Lc is the total length (round trip) of the cable in meters,

- cV the desired voltage drop.

## Cables Section Calculation

To calculate the cross-section of the cables that start from or from the photovoltaic panel and connect to the Charge Controller we use the formula:

$$S = K * I * Lc / cV$$

In our case: the panels are 4; current 5.95 A; voltage 42 V; length 10 m; voltage drop 0.48 V( 2% of 24 V), system voltage 24 Volt. The result is: 8.9 mm². We will use 10 mm² cables type FG21.

$$S = K * I * Lc / cV = 0.0175*12.3*20/0.48 = 8.96 \text{ mm}^2.$$

Where:

- K= resistivity:0.0175;

- I = maximum current (Isc) circulating, in our case 6.15 x 2=12.3 A as the panels are in parallel;

- L= cable length (10×2=20 m);

- cV= voltage drop (estimated at 0.48 V= 2% of 24 V).

Clearly if the length varies, as a consequence, the section varies. Example if the length was 5 meters, the section would be: S = K * I * Lc/ cV= 0.0175*12.3*10/0.48= 4.48 mm$^2$.

As said before the voltage drop must be in the 2%, we see if the condition has been respected and we use this formula:

$$\Delta V\% = (((P_{max} * K*Lc)/(S*V \text{ at maximum power}^2))*100$$

- $\Delta V$= voltage drop;

- Pmax= maximum PV power;

- K ($\rho$)= resistivity;

- Lc= cable length;

- S= cable cross section;

- V = voltage at maximum power.

$\Delta V\% = ((1000*0.0175*20)/(10*42^2))*100=1.98\%$ *(less than 2% so it is ok).*

## Calculation Cross Section Cables from Battery to Inverter

If I have to calculate the cross-section of the cables (6 meters long between return and battery) starting from the battery (24 volts) to the Inverter (1000 Watt), and the voltage drop is 2%, I will have with the calculation program: the cable cross-section will be 9.27 mm² which I will bring to 16mm² type FG21.

The result is more or less the same if we use the formula:

$$S = K * I * Lc / cV = 0.0175*42(I=W/V=1000/24=42) *6/0.48 = 9.18 \text{ mm}^2.$$

# CHAPTER 15: Directions for Photovoltaic Panels

Thanks to a practical mathematical formula and a map of latitudes, you will learn how to calculate the optimal inclination both in summer and winter in your city.

## In which direction should the Panels be oriented?

When designing the photovoltaic system, it is essential to choose the side of the roof on which the panels are to be mounted. We therefore need to know on which side there is more solar radiation during the whole day in order to have the maximum possible energy production. Since the photovoltaic panels are more productive when the sun's rays are perpendicular to their surfaces, the best orientation is certainly the one directly to the SOUTH (azimuth corner).

In case it is not possible to install in a southerly direction or we have shading problems (such as a very tall tree) we can slightly vary the orientation of the panels.

In fact, it is useful to know that if we orient the panels outside the south direction, up to a maximum of 45° (south-east and south-west), the annual production will suffer a fairly small reduction (1-3%). The solar radiation that the photovoltaic panels receive will in fact be almost the same. However, if the panels are facing at an corner greater than 45° to the south, production will start to decrease considerably.

At 90° from the south direction (i.e. directly east and west), the decrease in production can be up to 30%. This decrease is due to the fact that the photovoltaic panels, during most of the day, are affected by weak and not perpendicular sun rays. These panels will certainly produce photovoltaic energy but to a lesser extent than panels oriented directly to the south.

## Optimal Tilt of Panels

First of all, it is good to say that when choosing the inclination of photovoltaic panels on the roof of a house we are bound to satisfy two fundamental needs: the need for energy production and the need for an aesthetically pleasing and long-lasting final result. The roof of a house already has its own inclination and slope so we have to look for a compromise between these two needs. Knowing the optimal angle for production purposes will allow us to choose the best compromise for our installation.

The optimal inclination that our photovoltaic panels should have is essentially influenced by two factors:

- the latitude of the geographical location where we want to install them.
- from the time of year when we need more energy

## How does latitude affect inclination?

As we said before, the more a photovoltaic panel is inclined perpendicularly to the sun's rays, the more it produces electricity. The maximum production of energy must be obtained at noon, when the sun reaches its maximum height on the horizon. We therefore need to know, over the course of the year, the maximum and minimum height of the sun at midday to know how many degrees our panels tilt. This is why two particular days of the year come to our aid.

As many people already know, there is one day of the year when we have fewer hours of light and one day of the year when we have more hours of light. These are the days of:

- winter solstice.
- summer solstice

where we have more hours of daylight, is the 20th or 21st of June, and the sun at noon is at its highest annual height; and the winter solstice, where we have fewer hours of daylight, is the 21st or 22nd of December, and the sun at noon is at its lowest annual height.

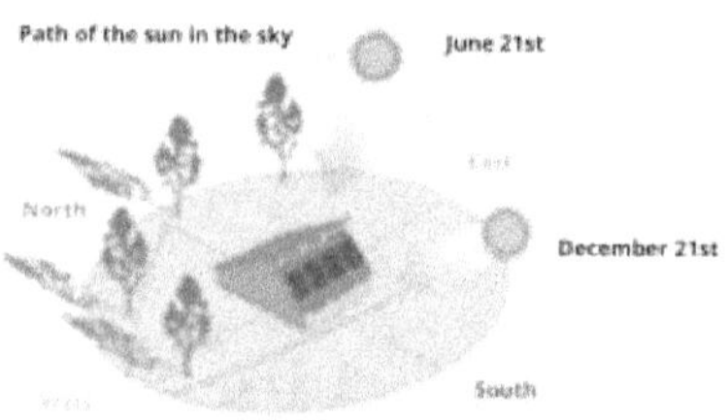


Geographical orientation and tilt angle can affect the energy efficiency of the system

Also depending on the latitude where we are, the maximum and minimum annual height of the sun changes at noon. In order to know the height of the sun that interests us during the summer and winter solstice we must identify how many degrees is the latitude of the place where we want to install the photovoltaic system. If, for example, we are in New York, the latitude will be about 40° (exactly 40° 43′).

**Calculating the maximum height of the Sun at midday**

Once the latitude has been identified, we must subtract 23° (current approximate terrestrial angle of inclination). The last mathematical operation is: 90 - (result obtained). In the case of New York, we will have the culmination point of the Sun at 73°, obtained from the calculation 90-(40-23). In order to obtain the maximum energy yield during the summer solstice, photovoltaic panels in New York must be facing SOUTH and tilted 17° (90-73). In this way at noon the panels will be perfectly perpendicular to the sun's rays.

## Calculating the minimum height of the Sun at noon

To calculate the height of the sun at midday during the winter solstice we must instead of subtracting, add 23°. The next mathematical operation remains unchanged. The result for New York is 90-(40+23) =27°. In order to obtain the maximum yield during the winter solstice, photovoltaic panels in New York must be facing SOUTH and tilted 63° (90-27°). In this way at noon the panels will be perfectly perpendicular to the sun's rays.

Calculations immediately show that a strong inclination of the panels with respect to the plane will favour a higher energy production during the winter period, while a slight inclination of the panels with respect to the plane will favour a higher energy production during the summer period.

Which inclination to choose for our system? Simple, the one that meets our energy needs.

If we are in New York and we need energy exclusively in the winter period it is good to choose an inclination between 55° and 63° (example 59°), while if we need energy exclusively in the summer period it is good to choose an inclination between 17° and 28° (example 24,5°). If, on the other hand, we need more during the summer period it is better to have an inclination between 17° and 40° (example 30°). If instead we have a greater need during the winter period it is good to have an inclination between 40° and 63° (example 53°).

If our energy needs in New York are constant throughout the year then we can opt for a middle way, such as 40°, always with orientation to the SOUTH.

# CHAPTER 16: The Cost of a 6 kW Photovoltaic System

Usually a 3-kW photovoltaic system is sufficient for a "standard" family of 3-4 people. If, however, you want to increase the production capacity of your home, a family, if they have space available on the roof of the house or in the garden, can opt for a 6-kW system. A 6-kW photovoltaic system realized on the classic pitched roof occupies about 50 square meters: with a free rectangle of 10 meters by 5 you can put panels for a production well above (double) the needs of a "standard" family.

## How much does a 6-kW system produce?

A 6-kilowatt (peak) power plant produces on average about 8,000 kWh/year of clean energy. In the south these 8,000 can reach about 9,000 kWh/year, although, of course, the actual yield of the system will depend on the climatic conditions of the installation site, the orientation and inclination of the photovoltaic modules. Inclination and optimal orientation of the panels at our latitudes are respectively: 30-35 degrees south. In addition, each system has a physiological drop in yield of about 1% per year and in the return plans it will be important to take this into account, even if in the long term it does not affect the economic return on investment.

Those who opt for a photovoltaic system of greater size than the classic 3 kW, therefore, can move towards a 6 kW which, in order to be fully exploited to its full potential, will have to be used as much as possible in self-consumption. In the case of a 6-kW system used to serve a family with "standard" consumption, it may be convenient to replace some gas appliances with electrical appliances. This will optimize the use of the system with a higher share of self-consumption.

But what is the cost of a 6-kW photovoltaic system?

Obviously a 6-kW system has a higher price, which can be better amortized by feeding the consumption of the house with electricity from your own system. How many dollars? Obviously, the answer is not univocal, nor definitive because it depends on the price of photovoltaic modules, inverters and the cost of the works for the realization of the system.

The first question in the mind of those who want to install a 6-kW photovoltaic system, instead of the 'traditional' 3 kW domestic one, is the cost: what is the cost of a 6-kW photovoltaic system? Can the higher cost be offset by the higher production?

The prices of turnkey installations no longer vary significantly.

As we said, however, what matters most in the photovoltaic investment is not so much the cost incurred, but the time of return on investment. To minimize the economic payback time of a home or business installation the principle is always and only one: self-consumption. If you plan from the point of view of self-consumption, obviously, the size of the system will have to be calibrated on the actual consumption of the family (or company). Among the electricity consumption of a family, can include, in addition to those for traditional household appliances, also the costs of heating, cooking or appliances for which methane gas is traditionally used.

In a nutshell: by replacing traditional gas appliances with new energy-saving electrical appliances, the higher cost of a larger system is offset by higher production.

## Cost of a 6 kW PV system

Today we're about $2,000 per kilowatt installed turnkey.

Providing an aluminium structure to fix the panels on a sloping pitched roof and all the accessories and components necessary to make the system work, the turnkey cost for a "standard" installation starts from about $ 14,000.

This is, of course, an indicative price that can be found in several price lists.

# CHAPTER 17: Maintenance of the PV System

Having a photovoltaic system can be a great way to save energy. But for this to work perfectly, allowing us to transform solar energy into electricity, it is necessary to periodically check the system to keep its productivity levels always high and ensure a longer life of its components. For a correct functioning of the photovoltaic system it is, therefore, necessary to take care of its maintenance to prevent a whole series of accidents, such as a fire, which can damage it or reduce its performance. It is therefore very important to know your photovoltaic system well in order to be able to act at its best and keep it efficient.

## What is Maintenance

First of all, to ensure proper operation of the panels, they must be kept as clean as possible. Dust, dust, soil, external elements, such as leaves or birds' nests, can be the cause of the malfunction of the system. The dirt that is deposited can inhibit the absorption of sunlight and thus reduce the accumulation of energy. Seasonal cleaning of solar panels is always recommended. Although rain can help to keep the panels clear, it is often not enough and a more incisive intervention is needed, which may require the use of a specific detergent or suitable tools. It is best not to use a sponge that is too abrasive when cleaning, as this could damage and scratch the structure of the panels. Water, soap and elbow oil are often sufficient to maintain the system at its best, but there are many kits on the market to make sure you use the right tools. Another important trick is to dry the modules well and avoid leaving halos, which can negatively affect the performance of the structure, although in a lesser way than dirt. To ensure the correct functioning of the system, it is always advisable to monitor the performance obtained. In this way, having the situation under control, we can immediately notice anomalies and faults.

The classic ordinary cleaning operations can sometimes not be easy at all. In case the solar panels are placed in places that are difficult to reach, as when they are placed on a roof for example, you can use special telescopic tools that help in cleaning and reaching the most difficult points. Although the best solution, especially for those who have no time or who want to be sure to have a job well done, is to turn to specialized companies. Usually these companies take care of the maintenance of the system at 360 degrees, dedicating themselves not only to cleaning, but also carrying out a visual check to detect any damage, monitor the performance of the system and make sure it operates in complete efficiency.

## What is the best time to clean

As we have already pointed out, it is good to plan a periodic cleaning of the photovoltaic systems. The best period to carry out maintenance is the time prior to the period of increased production. In simple terms, the most suitable time is the beginning of spring. The winter months, in fact, put a strain on the systems due to the strong temperature fluctuations, the cold temperatures, but also because of the snowfalls. The weight of the accumulated snow can, in fact, cause damage to the plant, which should be checked to monitor its proper functioning.

In addition to the cleaning of the solar panels, which as we have seen is an essential element, there are other very important precautions, which contribute to maintaining the perfect efficiency of the system. Let's see in detail the different interventions that must be carried out.

## Inverter control

Inverters are the heart of every system. They are the means by which solar energy is converted into electricity. In addition, they are responsible for monitoring the entire system and enable it to work at peak performance at all times. Usually they have a 10-year warranty and the monitoring, unless a fault occurs, should be done at the end of the 10-year period. The inverters are overhauled by specialist personnel, who monitor their correct operation and ensure their efficiency.

## Check cabling and electrical connections

This control should also be entrusted to specialized personnel, who will periodically check that the whole system is working properly and that there are no faults or problems related to electrical connections.

# Checking the performance of photovoltaic modules

The control of the production of solar panels is a very useful tool to monitor the efficiency of the photovoltaic system. In fact, a drop in production can indicate problems, such as a failure or the presence of dirt. But it may happen that the problem is attributable to damage to one of the cells, which in this case will have to be replaced. A thermal chamber is often used to check the modules, which immediately detects faults and malfunctions.

# Checking the antifreeze level

Checking the level of antifreeze, which helps prevent temperature changes during the winter months from damaging the photovoltaic modules, is also an important element to monitor. As we have seen, cold temperatures put stress on the system, which must be able to protect the cells from frost.

# Battery Control

To limit maintenance and increase the profitability of the system another element that could be considered is to add a lithium storage battery. This small trick allows you to make the system more efficient. A storage battery allows, as the word goes, to store unused solar energy and to spend it at times of the day when you are forced to use the power provided by the grid, in the evening and at night. In this way you have the opportunity to drastically reduce your bill costs and maximize the use of the photovoltaic system. In addition, with a storage system maintenance is reduced to a minimum, compared to a traditional battery.

# CHAPTER 18: Installation Panels

The first thing to do is to assess the condition of the roof, in order to establish the remaining life span of the current roof. The entire roof surface must be inspected very carefully in order to identify any points in need of repair. In addition, any problems with moisture or seepage should be investigated and dealt with accordingly. It is important to ensure a useful life of the roof of at least 20 years - the typical life of a photovoltaic system and government incentives - in order to avoid the cost of removing the photovoltaic system to carry out repairs on the roof. If the roof is made of eternit or contains asbestos, the installation of the photovoltaic panels is an excellent opportunity to proceed with the renovation of the roof, given the potential danger that such a roof poses to human health. In addition, an expert must assess the resistance of the roof to the weight of the panels and to strong winds that hit them.

## Installation of PV panels on flat roofs

First load the photovoltaic panels on the roof, taking care not to damage the roof covering. Then, you determine the east-west direction, along which the metal linear supports (stainless steel or galvanized, or aluminium) on whose rails the panels will be attached, which must be oriented to the south. Many types of structures are fixed by drilling holes in the roof and using dowels, but there are also pre-assembled frames that can be installed on the flat roof without drilling holes thanks to ballasts. In the first case, apply sealant around the holes to ensure a watertight seal and prevent moisture from entering the building. Finally, assemble the panels, which will then be raised with respect to the roof, and then proceed with the relevant electrical connections. The last step is the fine adjustment of the inclination of the panels: for each latitude, in fact, there is an optimal value that allows to maximize the solar radiation. (See Chapter 15).

## Installation of PV panels on pitched roofs

Photovoltaic panels can be fixed to pitched roofs - i.e. on pitched roofs (typically at an angle similar to the ideal angle to maximize solar radiation) - by means of special metal frames, suitably cut to size, which allow partial or total integration to the roof itself. These supports guarantee a high degree of safety against snow and wind loads, and high resistance to corrosion by atmospheric agents. The first phase of assembly consists in fixing the anchorage brackets and the relative uprights to the roof, by means of expansion bolts or chemical inserts, on which the actual panel-holder frame will be mounted, once the holes have been properly sealed. In this way, it is possible to make panel matrices consisting of a single row or many adjacent rows. In the latter case, each row of panels will be connected to the previous one by a suitable connection profile. Finally, the electrical connections of the system will be made.

## Installation of panels integrated in pitched roofs

A photovoltaic system, as well as being partially integrated on the roof in the way just described, can replace the roof itself, i.e. be "architecturally integrated", as they say in jargon. In this case, the first step is to lay a double waterproofing sheath on the "bare" roof on which the supporting frame of the photovoltaic panels will then be installed. As usual, we first proceed to fix the metal brackets with expansion plugs or chemical inserts, or similar systems of adequate strength. These brackets are then covered with a layer of bituminous sheath to prevent possible infiltration through the anchor screws. The uprights, which will support the frame itself, are then attached to the brackets. As screws, stainless steel screws are usually used, which resist deterioration over time, but instead special locking systems can be used to guarantee the customer against any attempts to steal the panels.

# Configuration of a structure for pitched roof PV systems

## Structure for PV systems 1kw - 4 modules

For the installation of a 1KW photovoltaic system on a pitched roof consisting of 4 modules, divided into two rows of 2 modules each, it is possible to adopt this structure:

- Bracket n° 8. We recommend the use of 2 brackets for each row of profile in roofs such as wood or concrete.
- Profile 32x32mm from 2,15 mt - 4 rows n° 4
- Central panel stop clamp n° 4
- Panel stop terminal clamp n° 8

## Structure for 2kw photovoltaic systems - 8 modules

For the installation of a 2KW photovoltaic system on a pitched roof consisting of 8 modules, divided into two rows of 4 modules each, it is possible to adopt this structure:

- Bracket n° 12. We recommend the use of 3 brackets for each row of profile in roofs such as wood or concrete.
- Profile 32x32mm from 2,60 mt - 4 rows n° 4

- Profile 32x32mm from 1,60 mt - 4 rows n° 4

- Profile joint n° 4

- Central panel stop clamp n° 12

- Panel stop terminal clamp n° 8

## Structure for 2kw photovoltaic systems - 8 modules

For the installation of a 2KW photovoltaic system on a pitched roof consisting of 8 modules, divided into two rows of 4 modules each, it is possible to adopt this structure:

- Bracket n° 12

- Profile 32x32mm from 2,60 mt - 4 rows n° 4

- Profile 32x32mm from 1,60 mt - 4 rows n° 4

- Profile joint n° 4

- Central panel stop clamp n° 12

- Panel stop terminal clamp n° 8

## Structure for 3kw photovoltaic systems - 12 modules

For the installation of a 3KW photovoltaic system on a pitched roof consisting of 12 modules divided into two rows of 6 modules each, it is possible to adopt this structure:

- Bracket n° 20. We recommend the use of 5 brackets for each row of profile, one every 1.30mt.
- Profile 32x32mm from 2,60 mt - 4 rows n° 8
- Profile 32x32mm from 1,00 mt - 4 rows n° 4
- Profile joint n° 8
- Central panel stop clamp n° 20
- Panel stop terminal clamp n° 8

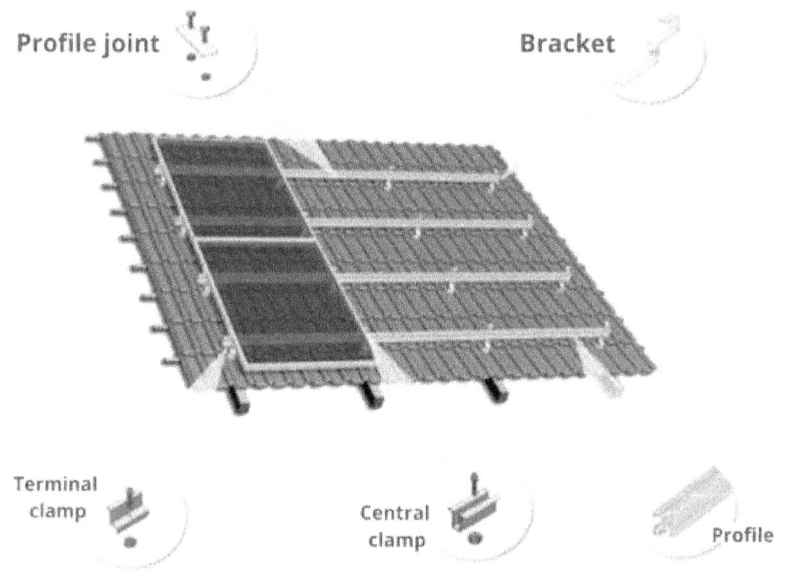

# Assembly Instructions

1) Identify the fixing point according to the design. Move the tiles and position the brackets by adjusting the height according to your needs.

2) Fix the profiles on the brackets

3) Connect the profiles with the joints

4) Position the first module on the profile by fixing the ends with the end clamps.

5) Place the next module next to the first one, fixing it with the central clamps.

6) Position the last module on the profile by fixing the ends with the end clamps.

# CONCLUSION

Thank you for coming all the way to the end of this book, we hope it was informative and able to provide you with all the tools you need to achieve your goals, whatever they may be.

This book has tried to bring all the important points to the fore so that you can get all the information you need on both how solar panels work and how to build a home PV system without having to deal with the negative effects.

You can also get all the benefits of the process by following the simple steps in the book.

I hope this guide will really help you achieve your goals.